Energy Management for the Metals Industry

Energy Management for the Metals Industry

Cynthia K. Belt
Energy Management Consultant
Metals Energy Management, LLC

CRC Press
Taylor & Francis Group
Boca Raton London New York

CRC Press is an imprint of the
Taylor & Francis Group, an **informa** business

CRC Press
Taylor & Francis Group
6000 Broken Sound Parkway NW, Suite 300
Boca Raton, FL 33487-2742

Printed on acid-free paper
Version Date: 20160915

International Standard Book Number-13: 978-1-4987-9925-6 (Paperback)

Library of Congress Cataloging-in-Publication Data

Names: Belt, Cynthia K., author.
Title: Energy management for the metals industry / Cynthia K. Belt.
Description: Boca Raton, FL : CRC Press, 2017. | Includes bibliographical references.
Identifiers: LCCN 2016035052| ISBN 9781498799256 (pbk. : alk. paper) | ISBN 9781498799294 (ebook)
Subjects: LCSH: Industries--Energy conservation. | Industries--Energy consumption. | Materials. | Industrial ecology.
Classification: LCC TJ163.3 .B437 2017 | DDC 620.1/10682--dc23
LC record available at https://lccn.loc.gov/2016035052

Visit the Taylor & Francis Web site at
http://www.taylorandfrancis.com

and the CRC Press Web site at
http://www.crcpress.com

Contents

Author

Cynthia K. Belt has developed and managed energy programs at Aleris International, Superior Industries, and Kaiser Aluminum. Within just a few years at each company, these energy programs improved energy intensity by 15%, saving the companies millions of dollars. She has published more than 15 papers in the area of energy management in the metals industry, has coedited several proceedings in energy and recycling, and has led workshops on energy management in the materials industry. Cynthia is a past chair of the Energy Committee within The Minerals, Metals, and Materials Society (TMS) and is currently on the TMS Board of Directors. She is involved in energy groups within TMS, American Foundry Society, and American Society of Mechanical Engineers. Cynthia earned her Bachelor of Science, Mechanical Engineering degree from Ohio Northern University and is a Six Sigma Black Belt.

Chapter 1

Introduction

1.1 Overview

This book is directed toward energy managers, engineers, supervisors, and managers working in the materials industry who want to reduce their energy usage. Its intent is to explain both general energy management methods and those areas specific to the metals and materials industry. Topics include

- Why implement energy management? Benefits and reasons for an energy management program. What is required to begin an energy program? Assigning an energy director, forming an energy team, making an energy policy, and finding resources.
- Defining and measuring performance. What should you measure? Understanding energy measurements using electricity, natural gas, coal, and so on. Understanding critical Key Performance Indicators. Setting baselines and benchmarking your plant and processes.
- Analyzing the data and identifying projects. Understanding your energy use with data analysis, system analysis, and technical assessments. Making a project

list. Improving the process. Setting goals and creating an action plan.
- Controlling the process and evaluating progress. Implementing a Project Action Plan. Team meetings. Tracking your energy performance. Recognizing achievements.

In each topic, specific information and examples for the material industry are highlighted, giving a more in-depth look at energy management within the materials industry.

The primary units in this book are metric (International System of Units [SI]); however, units used in the United States are also prominent. Please watch the units carefully because they can change depending on the source material of the data. Costs in this book mention United States dollars ($).

1.2 Purpose

Maybe your company has already started an energy program or maybe you are starting a new program. Maybe the purpose of your energy program is to be a more sustainable company or maybe it is to save money. For whatever reason, you want to learn more about energy management in the materials industry.

One can find many courses and training programs on energy management. But training for applications in the materials industry is limited. And frankly, we're different. While many plants are worried about whether to install light-emitting diode (LED) or T5 lighting, the energy used for lighting in metals plants is normally less than 5%, and for many plants it is less than 1% of the total energy used. Although much of the available information discusses energy used for HVAC (heating, ventilation, and air conditioning), many metals plants have limited heating and air conditioning. Instead, process heating is our most important area of energy use. Energy managers in our industry must not only understand the concept of energy management but also understand process heating.

The materials industry must use a large amount of energy to melt and heat treat metal. Energy costs are normally a major portion of our cost (labor and metal costs are also high). In the United States, the primary metal industry uses 8% of all energy used in manufacturing. Sadly, although energy use is high, energy management is not universal within the industry. Counting both internal programs and external ones such as the utility supplier, equipment manufacturers, and federal and state programs, only 66% of U.S. primary metals companies participate in any energy management activity (Belt 2013). Energy efficiency is not as good as it could be. It has been estimated that energy use in the iron and steel industry can be reduced by 39% with commercially available technologies and another 24% with new and innovative technologies (DOE AMO 2015). For aluminum it is estimated to be 33% savings with current technologies and another 30% with new technologies (DOE AMO 2016).

This book will allow you to

1. Understand the steps of energy management
2. Understand the special case of energy management in the materials industry
3. Obtain energy saving ideas

POTENTIAL PITFALL!

In Potential Pitfalls, I will let you know some of my mistakes so you can avoid making them.

1.3 Why Energy Management?

Energy is a controllable expense. Reducing energy costs improves your bottom line and energy efficiency is valuable

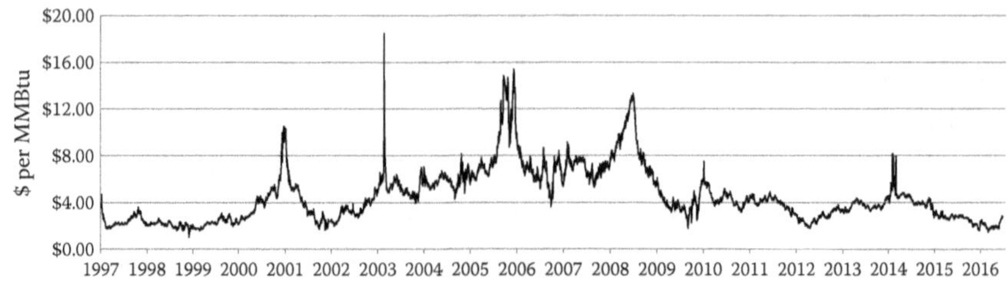

Figure 1.1 Henry hub natural gas prices 1997–2016.

to your business. Energy costs are typically 5%–7% of a company's sales in the casting industry (Eppich and Naranjo 2007) and can be higher in primary metals production.

Energy markets are undergoing significant structural changes. Long-term energy costs and availability are volatile. Will electric prices go up with increased renewables such as wind and solar? Will increased environmental issues require reductions in energy use? Will fracking of natural gas and oils be curtailed? Will natural gas be used in more sectors such as transportation, and will electricity production and prices increase? Will global warming concerns increase regulation? The 2015 United Nations Climate Change Conference has made energy management programs more critical, but how it affects every country and industry is still to be decided.

Energy prices will rise and fall (see Figure 1.1) (EIA 2016), but overall energy prices are expected to rise. We have little control of energy prices, but we can control energy use. Some of the benefits to be gained from energy efficiency are outlined in the following subsections.

1.3.1 Energy Savings vs. Increased Sales

Assuming an average 2.4% operating profit (average in 2005 for the metal casting industry) (Eppich and Naranjo 2007), it would take $1,000,000 in new sales to generate $24,000 of operating profit. Which is easier: finding $1,000,000 in new sales, or saving $24,000 in energy costs? It is difficult to persuade upper management on this, but capital should go for both new business and reducing costs on all products with projects such as energy efficiency.

1.3.2 Energy vs. Throughput vs. Melt Loss

Better energy efficiency normally means better throughput, melt loss, quality, maintenance, and delivery. These qualities are interconnected. Specifically in the metals industry, energy, melt loss, and throughput are a connected triangle. Changing one frequently changes another for better or worse. It's important to optimize the system.

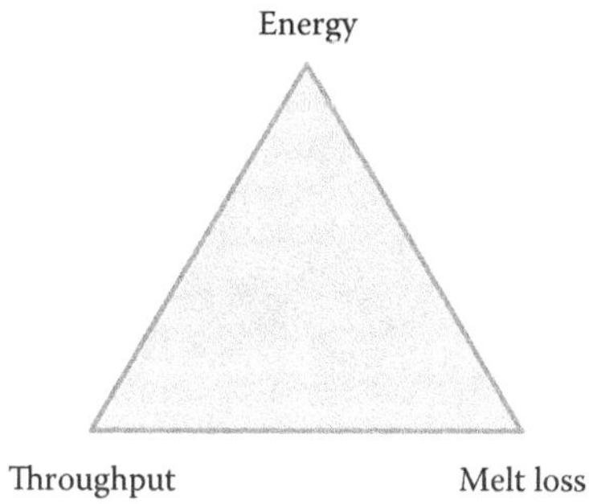

1.3.3 Environmental Factors

Government regulations in the United States require monitoring energy use. Energy is even more regulated in many other countries. If tracking and reductions aren't regulated now, they will be required in the future.

Almost every energy used in the metals industry means formation of greenhouse gases such as carbon dioxide (CO_2). This comes from natural gas, coal, and oil. Unless you are using 100% non-fossil fuel electricity (wind, hydro, solar), you are forming CO_2. Carbon credits or cap and trade may be implemented in the future in the United States and is common in other countries. You may or may not believe in global warming and climate change, but many people in science and government do. The good part is that energy programs help the environment *while* saving money!

1.3.4 Good Neighbor

Energy efficiency makes operations competitive and economically sustainable. Sustainability goals for the company should include energy management. Showing improved energy efficiency improves relationships with neighbors, municipalities, and shareholders and builds a history of environmental stewardship. Many times, it enables third-party recognition of your achievements.

1.4 The Energy Management Program

1.4.1 What Is an Energy Management Program?

Energy management sets up a program to conserve energy usage and improve energy efficiency. This program involves continuous improvement and is not a one-time project. Energy management can include energy purchasing, but this normally resides in a different department.

- *Energy conservation.* Conservation is mostly behavior oriented. Usage is reduced by optimizing the process so energy is applied only when it is needed (examples include reduced hold times, idling programs, motion sensors).
- *Energy efficiency.* Efficiency is mostly technology oriented. This is using the best technology or controls to reduce energy use (examples include regenerative burners, stack melt furnaces, combined heat and power units, LED lighting).

These terms are sometimes confused. Some projects accomplish both.

Older energy management practices need to be updated. Do any of these apply to your company?

1. Decentralized versus central corporate program. Sometimes plants or departments work to improve energy efficiency but the lessons aren't transferred to other areas.
2. Project-focused versus continuous improvement. New energy efficient furnaces are installed but energy use isn't tracked. Over time, the gains are lost.
3. Reactive versus proactive. When energy prices go up, energy projects are considered and then dropped when prices fall.
4. Short-term versus strategic. Energy projects may require downtime. When business is good, the company has money to make improvements but can't take the downtime. When business is down, no money is available even though time is available.
5. Limited by little accountability for energy use. Departments are not charged for energy use or energy

use is based on a set percentage, not real numbers. The department manager has no incentive to improve energy efficiency.

6. Based on scarce information on how to improve. Energy projects require training and research. When no one department is assigned this work, no one builds knowledge on potential energy-saving projects.

1.4.2 Myths

There are many myths about energy programs. Some of these are

1. "Only big companies can do it." Successful energy management is possible regardless of the size of your operation.
2. "Only plants with new equipment or large capital budgets can be energy efficient." Energy projects can address

capital or process changes. Some low-cost process changes can easily mean more savings and a better return on investment (ROI).

3. "We don't have enough staff." Part-time management programs work.
4. "It's so difficult we would need to hire outside experts." Lots of help is available online, with vendors, and with government programs.
5. "We are already doing everything we can." If you don't have a formal program, you aren't doing everything.
6. "Everybody in our company manages energy." This translates as nobody manages energy. In one real situation, a plant manager mentioned at the end of a staff meeting that natural gas prices had recently more than doubled. He mentioned that everyone should "think about" opportunities to save energy. Of course, nothing was done until finally a team leader was appointed to push on savings.

1.4.3 Good Energy Programs

What are the qualities of a good energy program? These may include

1. Energy management is a priority.
2. It has the support of senior management.
3. Responsibility is assigned with an energy leader, site energy champions, and performance goals.
4. It follows an Energy Management System; this is an established procedure and process for managing energy use.
5. It looks beyond first cost to total cost of ownership (TCO). For instance, energy costs over the life of a motor may be 10–90 times the purchase price.
6. It looks for both capital and behavior opportunities.
7. It focuses on continuous improvement.

1.5 Steps to Energy Management

Many different plans or methods are available to develop your energy management program. They all work, so pick what works for you and your company culture. In this book, we will be using the last version (Energy Management Plan).

Plan–Do–Check–Act (PDCA) (Deming Circle or Shewhart Circle shown in Figure 1.2) is a management method for continuous improvement. This plan has been around for years and it works.

Six Sigma DMAIC (Define–Measure–Analyze–Improve–Control, shown in Figure 1.3) is used by a team to improve a process. Many of these same steps will work in energy management.

The *ENERGY STAR plan* (Figure 1.4) (ENERGY STAR 2016a) is specifically for energy management. Information is available online (www.energystar.gov) on these steps used by many companies.

The *Energy Management Plan* (Figure 1.5) specifically applies to energy management but uses many of the steps

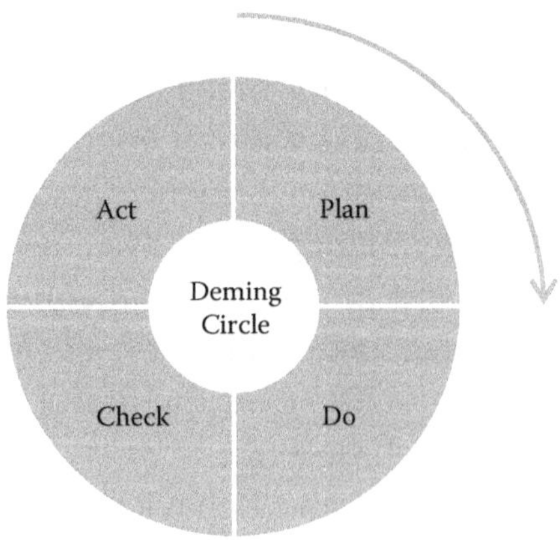

Figure 1.2 Deming Circle.

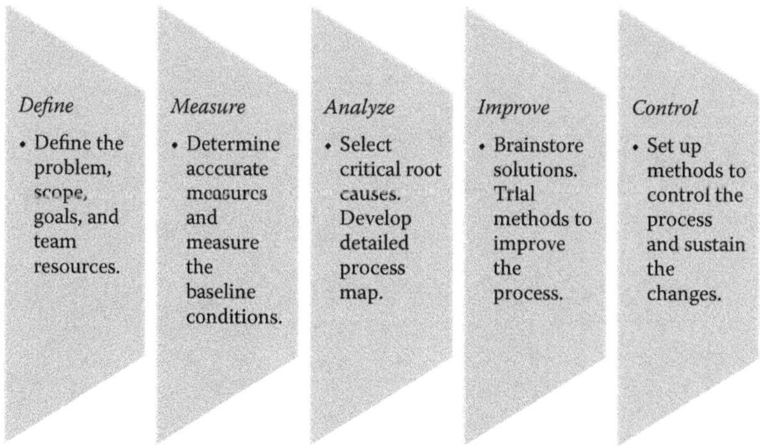

Figure 1.3 DMAIC process.

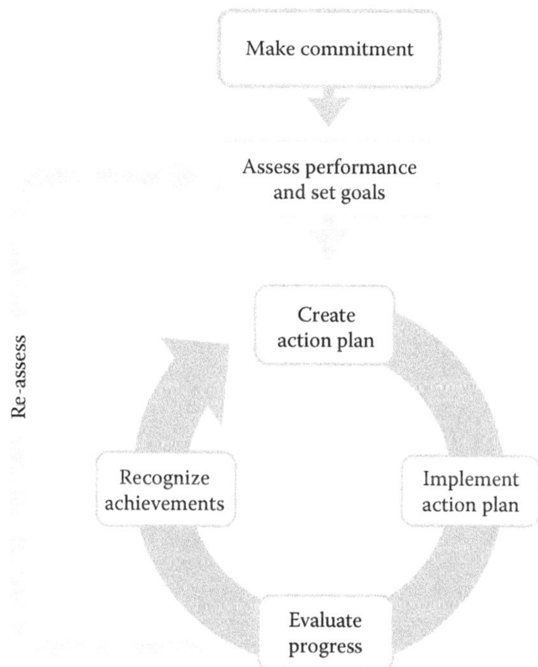

Figure 1.4 ENERGY STAR plan.

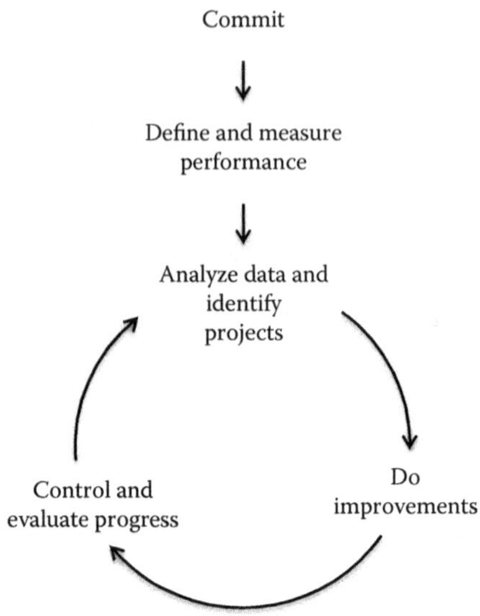

Figure 1.5 Energy Management Plan.

from these other programs. This is the one we will use in this book. The steps are

1. Commit
2. Define and measure performance
3. Analyze data and identify projects
4. Do improvements
5. Control and evaluate progress

We will go into detail on each step in the chapters that follow.

Chapter 2

Commit to the Program

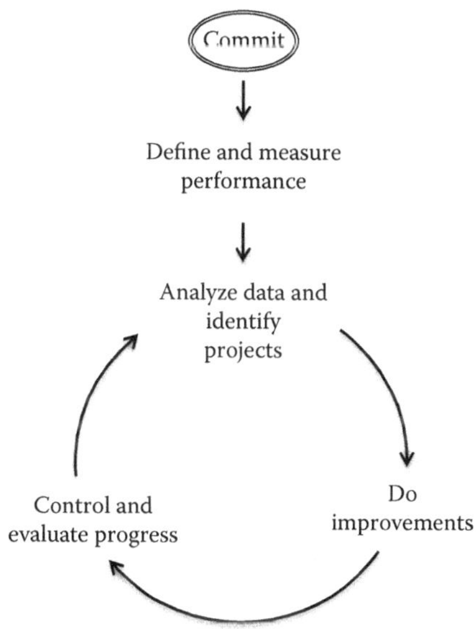

You've decided to start an energy program. There are steps you need to take to commit to this program and establish an energy efficiency initiative. These include designating an energy manager, pulling together an energy team, and composing an energy policy.

2.1 Energy Manager

The energy manager can be called energy director, energy engineer, energy manager, energy guru, energy czar, gas goddess, or whatever you like, but assign someone to run the program.

This is a full-time or part-time position, depending on the size of organization. If saving 1% of the company energy costs can pay for a yearly salary (more should be saved!), a full-time manager is recommended. It doesn't always need to be one person but can be a shared responsibility based on subject, duties, or regions. But make sure everyone understands his or her responsibilities. Responsibilities may include

1. "Owns" energy
2. Go-to person for all energy issues
3. Builds awareness among employees
4. Sets goals
5. Coordinates energy team efforts
6. Tracks progress
7. Helps to analyze data for trials or to understand developing issues
8. Investigates potential energy projects
9. External information gathering—keeps aware of new technology, energy tax grants, energy costs, and ideas from other energy management programs

10. Secures needed resources
11. Reports results and successes

POTENTIAL PITFALLS!

The energy manager requires constant training. Process heating and energy management are the most important, but proficiency in other areas is needed as well.

The energy manager needs to work well with other departments. The data is held by many people in accounting, maintenance, and operations. Most projects are run by other people in the company. The energy manager needs to persuade people to give him or her the data or to do the project work.

The energy manager needs to be enthusiastic AND persistent. Many of these projects take years to accomplish. The energy manager needs to keep up the push for improvements.

2.2 Energy Team

Rarely can the energy manager do all the work required for an energy program. Forming a team helps spread the work out, transfers the knowledge, and improves buy-in by different plants and departments. Is an energy team absolutely required? No, but an energy team absolutely helps.

The energy manager should lead the team. Although not an active member, having a team champion who is a senior manager can help motivate team members and help with any issues in prioritizing energy work external to the team. The team can include people from Operations, Maintenance, Engineering, Environmental, and Purchasing. If you have multiple sites, include a representative from each site to share successful practices. For larger companies a team may be needed for the corporation and another team for each plant. Consider changing team members over time—enthusiasm counts!

POTENTIAL PITFALLS!

Not enough time. The team member's supervisor needs to understand and agree to time spent for meetings and work done to improve energy.

Lack of knowledge. Consider sending team members to pertinent training. Maybe designate one member to be the expert on process heating and another to be an expert on compressed air. Can internal training be developed to train new members?

Wrong position. Sometimes someone is assigned who doesn't have the authority to make any changes or works in the wrong department. Perhaps another person can be added to run projects?

Cynicism. A member may believe this is just another short-term project that will eventually fail. This may take time and experience with successes to overcome. Having upper management attend several meetings may help. Or, another person may be better for the team.

2.3 Energy Policy

Many companies have a formal energy policy. An energy policy can

1. Formalize commitment to manage energy
2. Provide authority to implement energy initiatives
3. Establish a document to share with employees, customers, and other key stakeholders
4. Become a basis for building employee support such as allocation of staff and resources
5. Establish a focus and a foundation for setting performance goals
6. Engage upper management and show you mean business
7. Allow your company to achieve continuous improvement with energy and financial savings

An energy policy is recommended but not critical. There are successful energy programs without an energy policy.

The energy policy could also be a part of Environmental Health and Safety policies. The potential format could include

Purpose Statement: What are you trying to accomplish? What is the purpose or value of this work?
Action Statement: How are you going to reach goals? Include a clear understanding of intentions.
Value Statement: Why are we doing this? What is the higher purpose?

The example policies that follow are portions of good energy policies available on the Internet. They should not be copied!

XXX ENERGY POLICY

INTRODUCTION (DAVISCO 2016)

The objectives of this policy are to improve energy consumption efficiency, reduce cost, optimize capital investment for energy efficiency, reduce environmental and greenhouse gas emissions, and conserve natural resources.

POLICY

XXX will promote the efficient use of energy to produce and deliver products and services to its customers.

STATEMENT POLICY GUIDELINES

The following steps should be pursued to support the policy:

- Improve energy efficiency by establishing and implementing effective energy management programs worldwide that support manufacturing capabilities while providing a safe and comfortable work environment.

- Emphasize energy efficiency as a factor in product development and in process and facility design.
- Secure adequate and reliable energy supplies at the most advantageous rates and implement contingency plans to protect operations from energy supply interruptions.
- Encourage continuous energy conservation by employees in their work and personal activities.
- Drive further development of internal and external energy efficient and innovative technologies.
- Cooperate with governmental agencies and utility companies on energy programs.
- Support national governmental energy efficiency policies.

YYY ENERGY POLICY (BOEING 2014)

YYY is committed to

- Conducting operations in compliance with applicable environmental laws, regulations, and YYY policies and procedures.
- Preventing pollution by conserving energy and resources, recycling, reducing waste and pursuing other source reduction strategies.
- Continually improving our environmental management system.
- Working together with our stakeholders on activities that promote environmental protection and stewardship.

Chapter 3

Define and Measure Performance

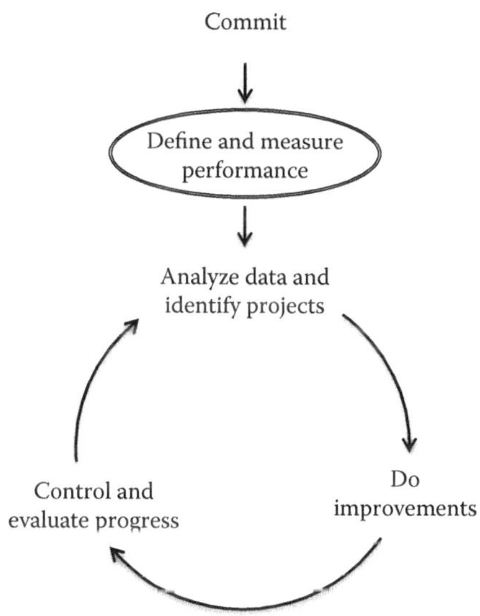

Measuring energy efficiency sounds easy, but there are a large number of details to be considered when calculating plant and furnace performance. We're going to discuss the general topic of data collection. The terms *power* and *energy* are frequently confused, so we will discuss the proper meanings. We go into detail about the energy types of electricity and fuels, including information on meters and heat content. Once we have the data, we need to use it to find the proper Key Performance Indicators (KPIs) for the plant and process. Finally, we need to set goals as a company and benchmarks for processes.

3.1 Data Collection

We've all heard the saying, "You can't manage what you don't measure." Subjective assessments such as "We need to improve" aren't enough. Real data is needed to find out where you are currently running and how much improvement is possible or how much you've improved over time.

First, determine the level of detail you want—plant, department, process, or furnace. Plant level is a minimum to measure progress, but most projects are identified and tracked at the furnace or process level.

POTENTIAL PITFALL!

No furnace is the same, even "identical" furnaces. Different product mixes, maintenance levels, products, and operators can all affect energy efficiency. Figure 3.1 shows four furnaces of the same type (manufacturer, burner systems, design, and dimensions) but with very different energy intensity (kWh/t or Btu/lb).

In the second example, a paint line had only one meter for the department but two different paint lines. Management knew there was a difference but had no data to substantiate the claim. Figure 3.2 shows the department performance. A new meter allowed the performance to be separated (Figure 3.3). One paint line was six times less efficient than the other! Given the data, an upgrade of the poorly performing line was cost justified and saved hundreds of thousands of dollars in a year (Peterson and Belt 2009).

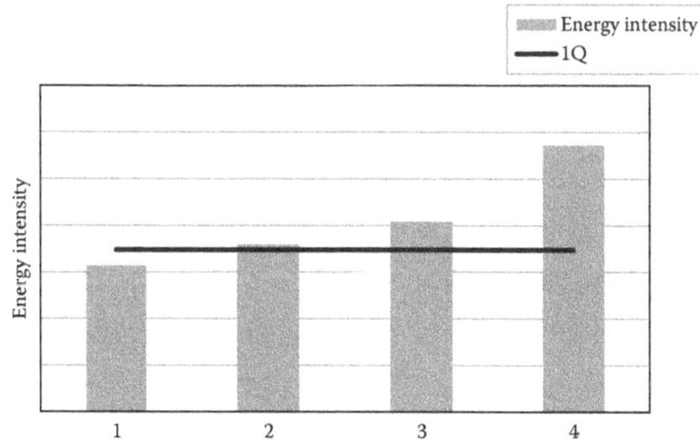

Figure 3.1 Energy intensity of identical dry hearth furnaces.

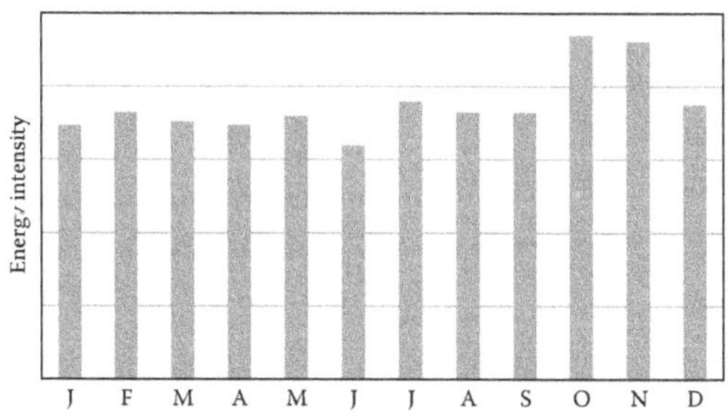

Figure 3.2 Energy intensity of the paint line department. (Adapted from R. Peterson and C. Belt, "Elements of an Energy Management Program," *JOM*, April 2009: 19–24. Copyright 2009 by The Minerals, Metals & Materials Society. Used with permission.)

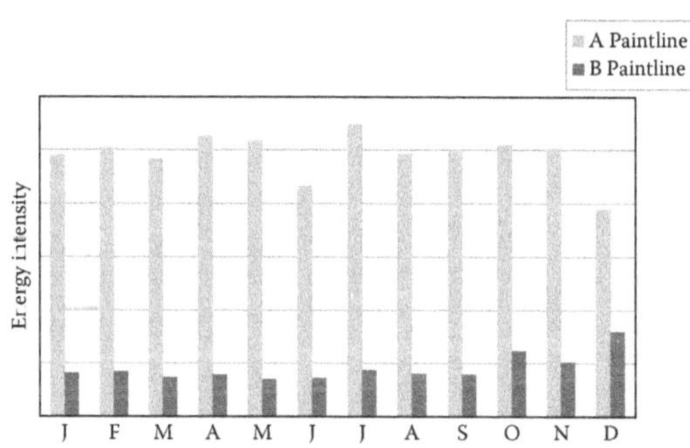

Figure 3.3 Energy intensity of two separate paint lines. (Adapted from R. Peterson and C. Belt, "Elements of an Energy Management Program," *JOM*, April 2009: 19–24. Copyright 2009 by The Minerals, Metals & Materials Society. Used with permission.)

Be prepared to collect data from multiple sources.

- *Energy data.* For natural gas, propane, diesel, fuel oils, and electricity these might include electronic or manual meter readings. The data may be collected by the network, maintenance, operations, or accounting departments, or the data may not be collected by anyone and you need to arrange for someone to collect data regularly. It is possible that the energy manager will need to make these readings directly.
- *Collect production numbers.* Production should include mass (in kilograms, pounds, or tons) and mass out (kilograms, pounds, or tons). Some industries use units of widgets out. For the metals industry, widgets out usually don't need to be collected. The number of ingots or castings is not as important as the mass.
- *Other data.* Other information is needed. Determine the fuel heat content from the fuel supplier. Energy costs are needed to understand the potential savings from projects. They include costs from bills, costs used in accounting (not necessarily the same), and forecasted energy costs. Other information such as plant square footage, operating hours, outside temperature, heating degree days (HDDs), and cooling degree days (CDDs) (not normally important in the metals industry) can help normalize for conditions over time and between plants.

  Average ambient temperature is important if the plants are heated. Cold metal and combustion air may also affect energy efficiency. Most accurate is an outdoor temperature reading at your plant, but it is easy to go online at sites such as www.wunderground.com and use the monthly average from the nearest airport.

  Heating degree days measures the amount of degrees and days that the ambient temperature is lower than a base temperature (normally 15.5°C, 18.5°C, or 65°F). A good source of these data is www.degreedays.net/.

Check for accuracy on all data. Every data source has some inherent inaccuracy. Determine if you can live with the inaccuracy you have or work to improve it. Meters are not exact, charge weight may be estimated, cast weight may be a calculated value, and so on.

POTENTIAL PITFALLS!

- *Don't believe everything you are told.* Check it! I once was told a department tracked gas usage for melt furnaces only. Then I was told the gas usage included hold furnaces. It turns out it included homogenizing furnaces too. Find the meters and trace the gas lines!
- *Analyze the data to see if it makes sense.* Data can show results that are too good. This may mean the wrong pressure and temperature factor is used or the meter is partially bypassed or is failing.
- *Dates and time may vary between sources.* I've found gas use was based on an accounting month while electricity was based on a billing month and production was based on a calendar month… all with different start and end dates. Daily numbers may be based on midnight or the start of a shift. Get consistent readings or correct your data.
- *Accounting numbers and real data may vary.* I've found natural gas use estimated based on a percentage of the plant gas use that was determined years ago. Accountants also tend not to correct erroneous data in a past month but will adjust a later month. That can skew your analysis.
- *Weights can be inaccurate.* Many times weights are estimated or scales haven't been calibrated recently. Sometimes you can live with the inaccuracies for monthly data as long as the error is consistent but watch this on specific tests.

3.2 Equipment List and Energy Profile

Conduct a facility walk-through and list major pieces of equipment. Record the pertinent information on large pieces of equipment. Make a simple Excel® or Word® file to record and compare data. The file is a good reference for these data and a way to compare performance and understand furnaces better.

Data can include

- Furnaces: Burner sizes, fire rate, burner control scheme, important furnace dimensions, capacity, process temperatures, charging and casting methods, types and condition of important furnace equipment, setpoints, and so on
- Motors: Horsepower (HP), utilization, nameplate efficiency, variable frequency drive (VFD)
- Other major energy users

POTENTIAL PITFALLS!

- *Physically check information* and don't rely on information from the offices. Sometimes I've found equipment still connected that management thought was removed. I've also found that the equipment on prints had actually been removed years ago.
- *Are furnace systems actually working?* For example, most plants will tell you the furnace pressure system is working when really the mechanism is stuck, the furnace pressure reading is wrong, or the setpoint is wrong.

A good way to understand which areas are most critical to address within the energy program is to develop an energy profile. This looks at energy use or costs per furnace, line, or department. Sometimes a profile is based on the types of energy systems such as process heating, compressed air, motors, and lighting.

Figure 3.4 shows one version that includes natural gas and electricity energy use summed over multiple aluminum processing plants (Belt 2012). This particular pie chart is based on energy usage. It doesn't take into account the costs of energy. Figure 3.5 is a single plant and is based on energy cost. Each method has its uses though energy costs mean more for potential energy cost savings.

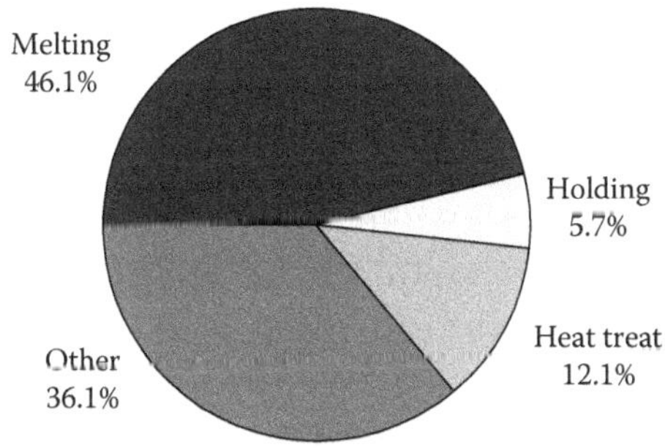

Figure 3.4 Aluminum processing plant energy usage. (Adapted from C. Belt, "Energy Opportunities in the Aluminum Processing Industry," *Energy Technology 2012: Carbon Dioxide Management and Other Technologies*: 57–64. Copyright 2012 by The Minerals, Metals & Materials Society. Used with permission.)

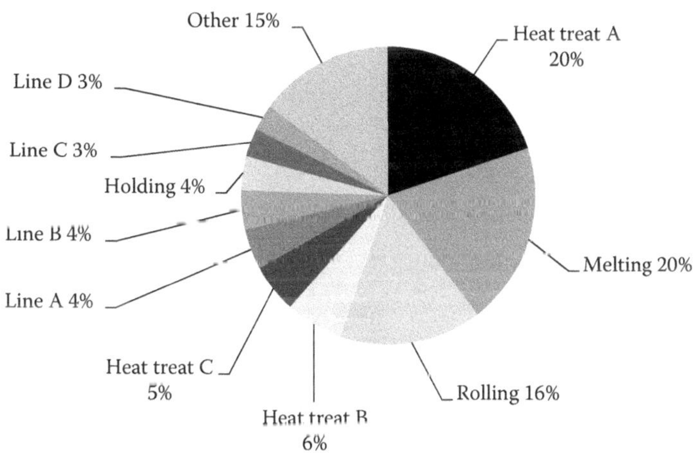

Figure 3.5 Plant energy costs.

3.3 Power and Energy

3.3.1 Power

Power is a rate, like the flow of water in liters per minute. Power is the rate at which energy is generated or used. The term *watt* doesn't sound like a rate but a watt is equivalent to one joule per second. It is named after the Scottish engineer James Watt.

Note that while most units are written with lowercase, the abbreviations for units named after people are capitalized such as watt (W) and joule (J, after James Prescott Joule). Symbols for multipliers of 1,000 or less such as kilo-, hecto-, or deca- are written in lowercase. Symbols for larger multipliers such as mega- and giga- are written in capitals. This means the symbols are written like W or kW or MW.

The lights in Figures 3.6 and 3.7 use roughly the stated instantaneous power.

Note:

1,000 W = 1 kW (kilowatt)

1 kW = 1.341 HP; 0.746 kW = 1 HP

Figure 3.6 A 60 W incandescent light bulb. (From https://upload.wikimedia.org/wikipedia/commons/3/3a/Gluehlampe_01_KMJ.jpg.)

Figure 3.7 1,000 W high pressure sodium (HPS) lights.

The blower motor shown in Figure 3.8 uses 7,500 watts, or 7.5 kW or 10 HP.

Note: 1,000,000 W = 1 MW (megawatt)

One of these large wind turbines shown in Figure 3.9 generates a maximum of 1.8 megawatts, or 1,800,000 watts.

Note: 1 kW = 3,412 Btu/hr

Figure 3.8 A 7.5 kW (10 HP) motor with blower.

Figure 3.9 A 1.8 MW wind turbine.

The burner shown in Figure 3.10 uses 4,700 kilowatts, or 16 million Btu/hr.

It can be called power or load or demand. Figure 3.11 shows power use over a month for a plant. The dips in use are due to weekend downtime.

Figure 3.10 A 4,700 kW (16 MMBtu/hr) burner.

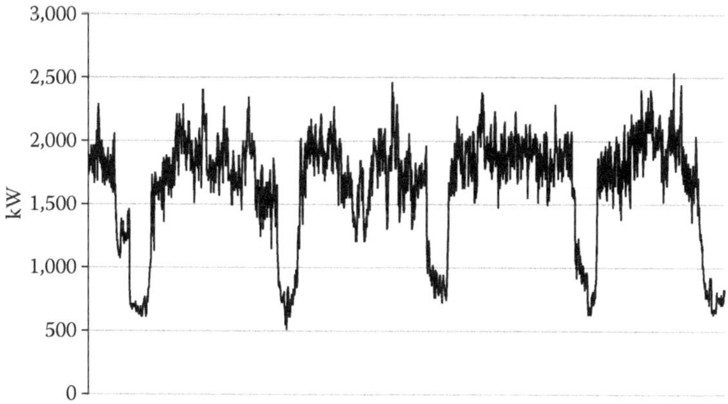

Figure 3.11 Electrical demand.

3.3.2 Energy

Energy is like volume or rate × time. Using our water example, this would be 10 liters per minute (power) × 1 minute = 10 liters (energy). You are paying the electric company for how many "liters" (kWh) you are consuming. Energy is a measure of fuel stored within something or used by something over a specific period of time. Think kWh, calorie, joule, or Btu. A cookie contains calories and a plant uses kilowatt-hours over a month.

$$\text{Energy} = \text{power} \times \text{time} \tag{3.1}$$

$$1{,}000 \text{ watts for } 1 \text{ hr} = 1 \text{ kW} \times 1 \text{ hr} = 1 \text{ kWh} \tag{3.2}$$

where
kW = kilowatt
kWh = kilowatt-hour

In the United States, most natural gas is purchased in Btu's. A Btu, or British thermal unit, is the energy required to raise 1 pound of water 1 degree Fahrenheit, or practically, the amount of energy in one matchstick. In the rest of the world, natural gas is typically purchased in joules. Energy analysis can be done in joules or kilowatt-hours or British thermal units.

For example, on a 4,700 kW (16 MMBtu/hr) burner:

$$4{,}700 \text{ kW} \times 8{,}760 \text{ hr/yr} \times 50\% \text{ utilization} = 20{,}586{,}000 \text{ kWh/yr} \quad (3.3)$$

$$4{,}700 \text{ kW} \times 3.6 \text{ MJ/kWh} \times 8{,}760 \text{ hr/yr} \times 50\% \text{ utilization} = 74{,}109{,}600 \text{ MJ/yr} \quad (3.4)$$

$$16 \text{ MMBtu/hr} \times 8{,}760 \text{ hr/yr} \times 50\% \text{ utilization} = 70{,}080 \text{ MMBtu/yr} \quad (3.5)$$

Power Conversions

1 W (watt) = 1 J/s (joule per second)
1,000 W = 1 kW (kilowatt)
1,000,000 W = 1 MW (megawatt)
1 kW = 1.341 HP (horsepower)
0.746 kW = 1 HP
1 kW = 3,412 Btu/hr (British thermal unit per hour)

Energy Conversions

1 kWh = 3,600,000 joules
1,000,000 J = 1 MJ (megajoules)

1 kWh = 3.6 MJ
1 kWh = 3,412 Btu
1 MJ = 947.8 Btu
100,000 Btu = 1 therm
1,000,000 Btu = 1 Dth (Dekatherm)
1,000,000 Btu = 1 MMBtu
1 MMBtu = 1 Dth
1 MMBtu = 1055.1 MJ
1 Btu = 252 calories (calories in food are actually kilocalories)

Another way of understanding the difference between energy and power is to look at a loaf of bread and dynamite. While the power of dynamite is greater, the energy in a loaf of bread is greater. Dynamite uses all the energy is a very short time while the energy in bread is used over a longer time. Table 3.1 shows the energy contained in some common items.

The important thing to remember is that power is a rate while energy is power used over a time interval.

Power = kW, Btu/hr, HP

Energy = kWh, joule, Btu, calorie

Table 3.1 Energy Content of Typical Goods

Energy Content	*Btu*	*kWh*	*MJ*
A match	1	0.0003	0.001
An apple	400	0.1	0.4
Pastry	2,000	0.6	2.1
Stick of dynamite	2,000	0.6	2.1
Loaf of bread	5,500	1.6	5.8
Pound of wood	6,500	1.9	6.9

3.4 Electricity

3.4.1 Understand Your Utility Bill

Every utility has a different cost structure and bill layout. Be prepared to see generation fees, transmission fees, distribution fees, taxes, credits, rental fees, and more. Carefully read your bill and make sure you read through your rate schedule. A rate schedule shows the details of the billing for a plant of your size.

Schedule a meeting with a representative of your utility company. The electricity representative can be your friend. He or she may have more experience with your company's energy usage than you do. Your utility can help you find opportunities for savings. Points to consider are

1. Make sure you have a copy of all your bills. It is not unusual for a single plant to have up to five different bills for different buildings or types of equipment.
2. Have your representative explain the bill and your rate schedule.
3. Understand how costs vary per time of day and season of the year.
4. Are there any ratchet costs? Sometimes a portion of the demand costs are based on the highest peak in the last 12 months.
5. Make sure you know what your power factor is for the plant.
6. Discuss if there are any other schedules or changes in your billing such as contracted demand that might save money.
7. Ask if there are any rental or lease fees (transformer or other). Does it make sense for these to be purchased instead?
8. Discuss what changes the plant can make to save money (control of peak, scheduling off-peak hours, etc.).
9. Ask about available plant assessments or copies of previous (and sometimes lost) assessments at your plant.

10. Ask about financial incentives offered by the utility. Many utilities will help fund projects such as lighting. Learn what the application process is, what equipment is covered, and how the project needs to be documented.

3.4.2 Electric Costs

Understand your electric costs. Different projects will reduce use (kWh), peak (kW), or both.

$/kWh = Add up "per kWh" costs

In the United States, charges vary from $0.02 to $0.18/kWh.

$/kW = Add up "per kW" costs

Demand charges (per kW) can be 25%–70% of the bill. Charges vary from no charge to $15.00/kW.

All-in $/kWh = Total bill cost ÷ kWh

POTENTIAL PITFALL!

Watch for differences in kWh used for a 28-day month versus a 31-day month.

3.4.2.1 Peak Demand

Instantaneous demand is the power use at any moment. Demand is averaged by your utility over 15- or a 30-minute period. Peak demand is the highest averaged point during the billing period (Figure 3.12).

To return to our analogy of water with liters and liters per minute: if you needed a high flow even for a short period, your pump and pipe must be sized for this high flow. In the same way, the electric utility company must be able to supply

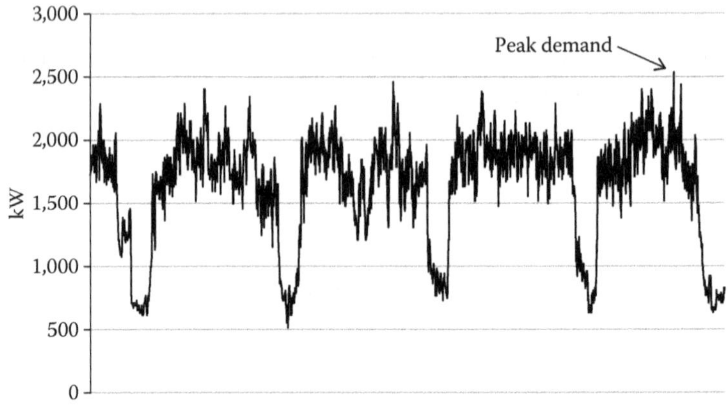

Figure 3.12 Peak demand during a month.

this power. The charge will be higher for companies with high demand because more generating capacity is needed even if it is used for only a short time. It may also require a larger transformer, wires, and substations.

POTENTIAL PITFALL!

Don't overreact to discussions of motor starting peaks. Remember, demand is the average over 15 or 30 minutes while high electricity use when a motor starts is only a few seconds.

3.4.2.1.1 Load Factor

Use the concept of load factor to understand how well you are handling peak demand. The goal is to have steady power use through the month. A poor load factor could mean the plant is running much of its high-energy users at one time. This costs the company more for peak demand.

In a story I heard, a company was paying for high peak demand. Investigation with the utility pinned the time of the peak to Saturdays, when the plant wasn't running! It turned out that a maintenance person was turning on all the air compressors at the same time on the weekend to make sure they were running properly. This resulted in a high peak. The practice was changed and saved the company a substantial amount.

$$\begin{aligned}\text{Load factor} &= \text{Average load} \div \text{Peak load} \times 100\% \\ &= (\text{Monthly kWh} \div \text{Hours per month} \\ &\quad \div \text{Peak demand}) \times 100\%\end{aligned} \tag{3.6}$$

Rule-of-thumb: Aim for 80% or better if running 7 days a week. Other goals are more appropriate for shorter work weeks (see Table 3.2).

POTENTIAL PITFALL!

A plant can have too good of a load factor. This can happen if the crew isn't idling during downtime correctly.

3.4.2.2 Power Factor

Plants with inductive loads (motors, transformers, etc.) cause the current to be out of phase with the voltage (reactive power) and is said to lag. The relationship is shown in what is called the power factor triangle (Figure 3.13). Real power

Table 3.2 Power Factor Goals Based on Production Hours

Days per Week	*Hours per Day*	*Hours per Week*	*Goal (%)*
7	24	168	80
5	24	120	60
5	16	80	40
5	8	40	20

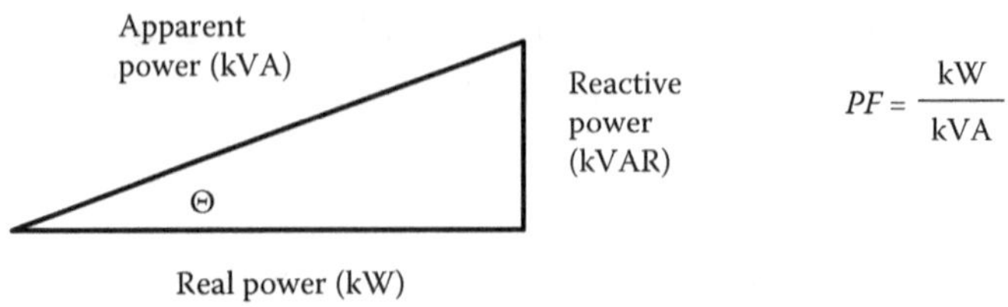

Figure 3.13 Power factor triangle.

does the work while apparent power flows to the circuit. When the power factor is poor, more current is pulled. Higher current requires larger wires and more energy loss. The power company charges more if the plant is running less than a certain value of power factor (normally 90% or 95%). Poor power factor is corrected by capacitors, typically in a capacitor bank.

This value is found directly on the bill, or instead you might see it as a penalty on the bill or it might be called reactive demand or as a kVAR fee. If it is not obvious, talk to your utility company.

Rule-of-thumb: Aim for slightly better than the requirement by the electrical company (90%–95%). There is no need to improve more than that. Note that not all utilities track or charge for power factor.

POTENTIAL PITFALL!

Many salespersons try to sell capacitor units that are added to motors to improve power factor. They may say they will reduce energy usage by 10% or more. Capacitors on a motor may save a small amount of energy, but more like 0.1%. Capacitor banks tend to be the better method to improve plant power factors. Plus, the salespersons will try to sell you these capacitor units even when your plant power factor is good. *It's a scam.*

3.4.3 Electric Metering

Use the electric data that is available from your utility and consider adding more metering. Monthly data is the minimum. More data will tell you more about your operation. Daily or 15-minute data is frequently available from the electrical company. When possible, add additional continuous metering inside your plant in departments and on major pieces of equipment.

3.5 Primary Fuels: Natural Gas, Oils, and Coal

3.5.1 Costs

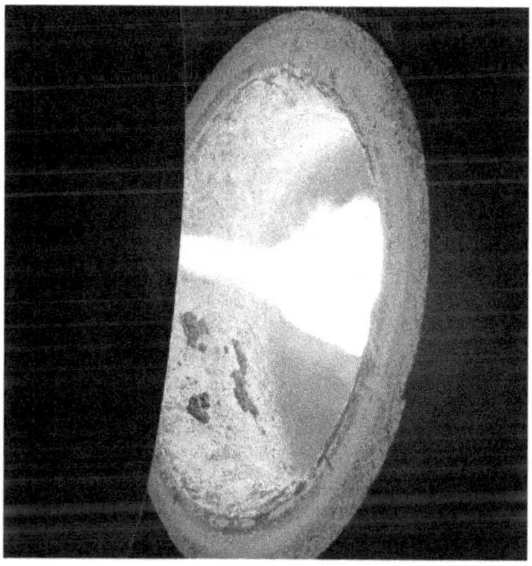

Collect costs for all your primary fuels. Understand your utility bill. Normally there is nothing but $/MJ (or other unit) but you may have special provisions. Note: hedging is outside the scope of this book!

3.5.2 Fuel Metering

Install required metering (monthly and real time). It is best is to have real time measurements to analyze opportunities. The minimum is to measure energy usage monthly to track progress. Some measurements may be from furnaces, departments, or a line of furnaces such as crucible/ladle heaters.

3.5.2.1 Other

The best energy tracking systems separately measure energy use for all but 5% of the gas use. I calculate "Plant Total" minus all measured sources and call this "Other." "Other" gas should be monitored too. This includes small sources, plant gas leaks, and meter errors.

POTENTIAL PITFALL!

"Other" gas can be used to see problems developing such as meter errors or gas leaks in pipes before the furnace meters. In Figure 3.14, the "Other" gas varied back and forth between positive and negative over two years, which shows normal meter errors. Suddenly the majority of the data showed positive and higher values of "Other." In this case, one of the furnace meters was failing.

Rule-of-thumb: Add meters for all processes using more than 1.5 MW (5 MMBtu/hr). My reasoning is as follows: 5 MMBtu/hr × 24 hr/day × 365 day/yr × $4.00/MMBtu × 66% utilization × 5% = $5,800. If a meter helps identify 5% savings, the meter is paid for in one year.

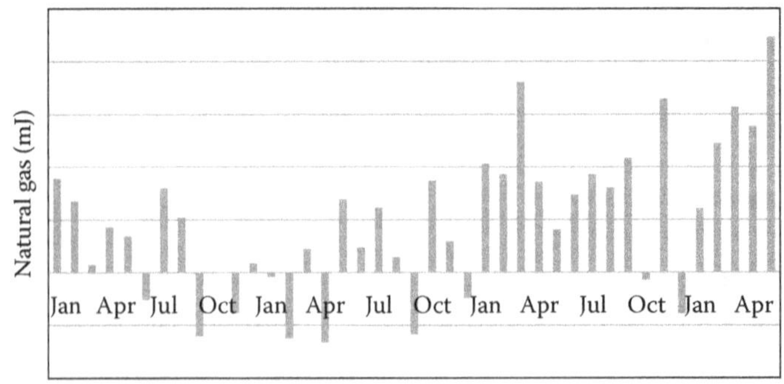

Figure 3.14 "Other" variation.

3.5.2.2 Plant Heat

Normally this isn't measured by one meter because the heaters are spread out over the plant. However, much of the time you can see plant heat use within the "Other" data. It's the difference between winter gas use and summer use (see Figure 3.15).

3.5.3 Estimates

If you don't have a meter, you can at least estimate use over the year to understand the potential for future projects. Add up the total of all metered gas and the estimates. Can you get close to total plant gas use?

Estimates are better than no data! Make estimates by multiplying the power by the utilization. You may need to monitor an operation for a while to understand the utilization. For instance, high fire may be on only 25% of the time even though the furnace is running 100% of the time. Or, the fire rate may run between high fire and low fire much of the time. Make your best educated guess. Use this value when estimating cost savings for a project in this area. This may also help you cost justify a gas meter for the area.

Table 3.3 shows a mixture of actual data along with estimated data. Summing these values and comparing them to the plant total gas usage is a good way to see if your estimates are close.

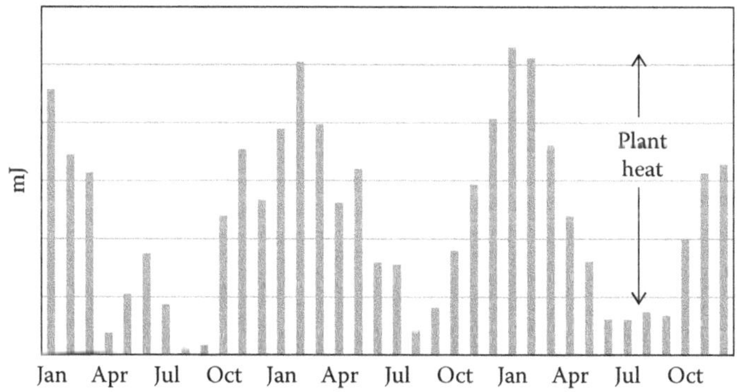

Figure 3.15 "Other" gas use.

Table 3.3 Plant Gas Usage Example

Gas Users	*Measured*		*Estimated*		*% of Plant*	*Tot. Burner Output (MMBtu/hr)*	*Qty.*	*Burner Rating (MMBtu/hr)*
	MMBtu/ Day	*Util. (%)*	*MMBtu/ Day*	*Util. (%)*				
Melting					**54**			
Melter 1	256	27			16	40.0	2	20.0
Melter 2	256	33			16	32.0	2	16.0
Melter 3	327	43			21	32.0	2	16.0
Holding					**20**			
Holder 1	75	35			5	9.0	3	3.0
Holder 2	123	47			8	11.0	2	5.5
Holder 3			115	40	7	12.0	2	6.0
Heat Treat					**17**			
HT 1			173	20	11	36.0	18	2.0

(Continued)

Table 3.3 (Continued) Plant Gas Usage Example

	Measured		*Estimated*					
Gas Users	*MMBtu/ Day*	*Util. (%)*	*MMBtu/ Day*	*Util. (%)*	*% of Plant*	*Tot. Burner Output (MMBtu/hr)*	*Qty.*	*Burner Rating (MMBtu/hr)*
HT 2	65	15			4	18.0	12	1.5
HT 3			32	10	2	13.5	9	1.5
Other					**9**			
Plant heat			85		5			
Crucibles			40	35	3	4.8	4	1.2
Small burners			10	10	1	4.0	2	2.0
Subtotal	1.102		455					
Sum				1,557 MMBtu				
Avg Plant Use				1,610 MMBtu				

3.6 Gas Meters

Given the large number of gas meters needed to measure energy use in our plants properly, a basic understanding of meters is needed.

3.6.1 Uncertainty

On all meters, there are four sources of uncertainty in the reading:

1. *Density.* The density of a gas varies with pressure and temperature. Some meters automatically compensate for this, some have an added controller unit to compensate, and some readings must be adjusted by you. See Section 3.6.3 on pressure and temperature factor.
2. *Piping.* The size of the pipe makes a difference for some meters. If the meter sensing unit is designed to be located in the middle of a 2-inch pipe, installing the unit on a 2 ½-inch pipe will induce error. In most cases laminar flow is needed for a correct reading. This means that a clear run of pipe of a certain number of pipe diameters is needed before and after the meter. This is the amount of space between the meter and an elbow, valve, or other obstruction divided by the pipe diameter. Check the manual or speak to your meter representative!
3. *Meter/transmitter.* Is the meter working? When was the meter last calibrated? Again, check the manual for the recommendation on calibration interval along with recommended preventative maintenance such as lubricating a turbine.
4. *Electronics.* Many meters have a set-up procedure. Go through each input to make sure it is correct (gas type, pipe diameter, etc.). Has the meter been zeroed? Is there an add-on unit for pressure and temperature compensation? Are the sensors installed correctly and is the reading a

reasonable number? If the data is sent to a network, is the right data going to the right place with the correct scale factor?

3.6.2 Types of Meters

Meters are not cheap, but they are important. Talk to your vendor about which meter is right in the particular application. Some of the types include

Orifice plate. An orifice plate is a plate installed between flanges in the pipe with a very specific diameter hole cut into it to provide a fixed obstruction. This obstruction causes a pressure drop so that measuring the pressure drop across the plate means the flow can be calculated (Figure 3.16).
Pros: Low cost.
Cons: Accuracy is poor at low flow. Pressure and temperature compensation must be added.

POTENTIAL PITFALL!

The plate should have a sharp edge on the upstream side but the plate is sometimes installed backwards. Many plates are marked on the upstream side. In dirty gas streams, buildup can occur on the edge of the orifice plate. Both of these issues detrimentally affect the results.

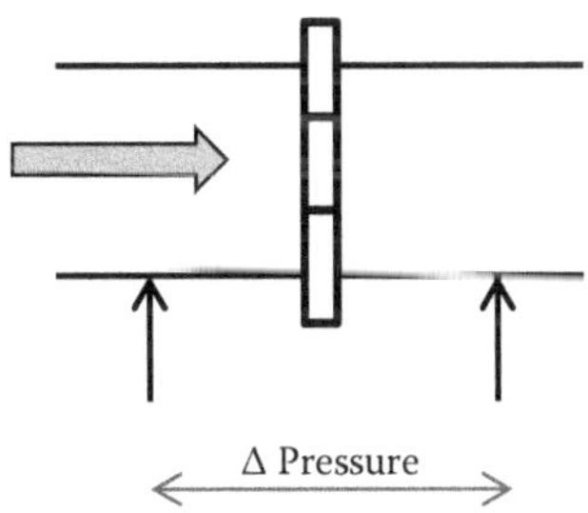

Figure 3.16 Orifice plate and pressure drop.

Turbine flowmeter. This meter measures the rate of spin of turbine vanes that are located in the flow (Figure 3.17).

Pros: Low cost. Accuracy is good at low flow (to a point).

Cons: Pressure and temperature compensation must be added.

POTENTIAL PITFALL!

Turbine meters can fail over time if they are not properly lubricated.

Thermal flowmeter. This meter uses a heated sensing element in the flow and an isolated element. The difference in temperature is proportional to the flow rate (Figure 3.18).

Pros: No pressure and temperature compensation is needed. Calibration is rarely needed.

Cons: These meters are more expensive.

Coriolis flowmeter. The fluid runs through a U-shaped tube. A precise vibration is added. Due to Coriolis forces, an

Figure 3.17 Turbine meter.

Figure 3.18 Thermal flowmeter.

additional vibration component is added so that the phase shift can be correlated to flow rate (Figure 3.19).

Pros: No pressure and temperature compensation is needed. It is very accurate at high and low flows.

Cons: These meters are more expensive.

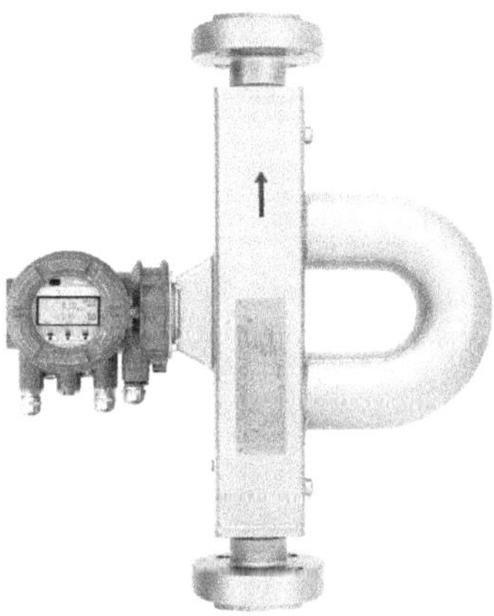

Figure 3.19 Coriolis flowmeter. (Used with permission from Yokogawa Corporation of America. http://www.yokogawa.com/us/solutions/products-platforms/field-instruments/flow-meters/coriolis-flow-meter/.)

3.6.3 Pressure and Temperature Factors

While many new meters are pressure and temperature compensated, some older or less expensive meters still require a correction factor. Check the manual. In most cases, if it is a turbine meter or an orifice plate it needs to be corrected.

Higher pressure means a higher correction factor (more mass within the volume with higher pressure). To calculate the pressure factor, read the gas pressure close to the gauge so pressure drops from regulators or piping are not included. If possible, use a recently calibrated or new pressure gauge.

Higher temperature means a lower correction factor (less mass within the volume with higher temperature). Read the temperature close to the gauge. As a shortcut, you can normally read the temperature of the pipe.

In pounds per square inch gauge (psig) and Fahrenheit units:

$$PF = (14.7 + \mathrm{P}) \div 14.7 \tag{3.7}$$

$$TF = (460 + 60) \div (460 + T) \tag{3.8}$$

$$CF = PF \times TF \tag{3.9}$$

$$\begin{aligned}\text{Actual flow} = (&\text{Meter reading at end}\\ &-\text{Meter reading at start}) \times CF\end{aligned} \tag{3.10}$$

where

PF = pressure factor
P = line pressure (psig)
TF = temperature factor
T = line temperature (°F)
CF = correction factor

Example for a gas line at 25 psig at 80°F:

$$PF = (14.7 + 25)\ \text{psi} \div 14.7\ \text{psi} = 2.70 \tag{3.11}$$

$$TF = (460 + 60)°F \div (460 + 80)°F = 0.96 \tag{3.12}$$

$$CF = 2.7 \times 0.96 = 2.60 \tag{3.13}$$

$$\text{Actual flow} = (140,326 - 134,980) \times 2.60 = 13,900 \text{ cf} \tag{3.14}$$

Similar correction factors are available for metric units. Read the meter manual to be sure these are the right formulas. Some meters are adjusted to other temperatures or pressures.

POTENTIAL PITFALLS!

Don't assume the gas readings are correct even with years of readings. I've found many instances of meters that were sticking, wrong pressure and temperature compensation factors, metric versus imperial errors, meter installed on the wrong pipe size, incorrect piping runs, and mix-ups in the transfer of electronic data.

There are times that the pressure or temperature changes. For example, gas lines that might cool off in the rain or heat up when nearby furnaces are running. If you don't have automatic compensation, the best you can do is an average.

Check the meter to make sure the decimal is in the right place (cf, ccf, or mcf) (cubic feet, 100 cubic feet, or 1,000 cubic feet).

Many meters have correction units. Physically check these too. I have seen pressure gauges improperly installed, thermocouples hanging in the air, and meter readings taken from the uncompensated gauge instead of the compensated gauge.

Periodic checks and calibrations are needed for many meters. Meters can be partially bypassed by an accidently open pipe run. And meters fail. As they fail, a process may look better. For example, the soaking pit energy intensity in Figure 3.20 appeared to improve. Eventually it even showed performance that was better than "perfect." Actually, the meter was failing and these measurements were incorrect.

Do a reality check. Even if you think you've checked everything, do the numbers appear reasonable?

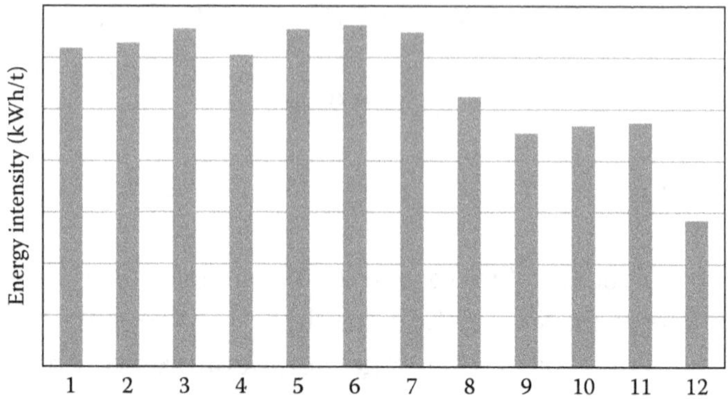

Figure 3.20 Soaking pit energy intensity. (Adapted from R. Peterson and C. Belt, "Elements of an Energy Management Program," *JOM*, April 2009: 19–24. Copyright 2009 by The Minerals, Metals & Materials Society. Used with permission.)

3.6.4 Heat Content

This is the amount of energy within a unit of fuel. The two measurements of heat content have several names depending on country or engineering, scientific, or other viewpoint. It all means the same.

Higher Heating Value (HHV) or Gross Heating Value (GHV) or Gross Calorific Value (GCV) or Higher Calorific Value (HCV) is the amount of heat obtained from a unit of fuel from perfect combustion of fuel and air starting and ending at a certain reference temperature. This includes the heat contained in the water vapor. It takes into account the latent heat of vaporization of water in the combustion products.

Lower Heating Value (LHV) or Net Heating Value (NHV) or Net Calorific Value (NCV), or Lower Calorific Value (LCV) is the amount of heat obtained assuming that the latent heat in water vapor is not recovered.

HHV is used in most cases in the United States. LHV tends to be used in Europe. The general conversion used is HHV = 1.11 × LHV, though the actual conversion varies based on chemical content of the natural gas. Neither method is necessarily right or wrong, but it is best to use the same method throughout the company. With an 11% difference, comparisons can be skewed because the difference.

Average heat content (HHV) for some fuels:

1. Natural gas: 1,025 Btu/ft^3 (38.2 MJ/m^3) or 10.6 kWh/m^3)
2. Landfill gas: 540 Btu/ft^3 (20.1 MJ/m^3 or 5.6 kWh/m^3)
3. Crude oil: 5,800,000 Btu/barrel (or 42 gallons) (551.5 MJ/L)
4. Gasoline: 125,000 Btu/gallon (34.8 MJ/L)
5. Diesel: 139,000 Btu/gallon (38.7 MJ/L)
6. Residual fuel oil: 6,287,000 Btu/barrel (597.8 MJ/L)
7. Coal: 19,858,000 Btu/short ton (23,095 MJ/t)

Coal has the largest variation in heat content. It can vary from 8,000,000 Btu/short ton (9,300 MJ/t) for lignite to 30,000,000 Btu/short ton (34,900 MJ/t) for anthracite (Bowen and Irwin 2008). Knowing the heat content of your coal is critical.

For natural gas, the heat content varies by the mix of gases (methane, ethane, propane, etc.). Heat content can vary from 900 to 1,150 Btu/cf but 1,025 is a typical value. It should be on your energy bill, but if not check with your natural gas or other energy supplier!

For natural gas, meters read cubic feet (or cubic meters). You need to multiply this by the heat content to get the amount of energy it contains. As an example:

$$1,000 \text{ mcf} \times 1,000 \text{ cf/mcf} \times 1,025 \text{ Btu/cf}$$
$$= 1,025,000,000 \text{ Btu} = 1,025 \text{ MMBtu} \tag{3.15}$$

$$28,300 \text{ m}^3 \times 38.2 \text{ MJ/m}^3 = 1,081,000 \text{ MJ} \tag{3.16}$$

$$28,300 \text{ m}^3 \times 10.6 \text{ kWh/m}^3 = 3,000 \text{ kWh} \tag{3.17}$$

POTENTIAL PITFALLS!

- *Heat content may vary over time.* Get readings from the company for several months or in different seasons to check.
- *Most furnaces do not utilize the energy within water vapor.* Although HHV is used in the United States, approximately 11% of the value is not available to heat the furnace.

Volume Conversions

100 cubic feet = 1 ccf
1,000 cf = 1 mcf

3.7 Secondary Fuels and Gases

Consider tracking propane, diesel, gasoline, and so on. Energy projects can be run to reduce internal or external transportation costs. Projects can include idling practices, reduced internal transports, and alternative fuels.

In our industry, consider gases such as oxygen, nitrogen, and argon. Water is also frequently tracked. Some companies even track airline flights and business car travel. Evaluate the percentage of the total energy costs, the potential for savings, or whether this is a part of an overall sustainability program before you add these. If your program is just starting, you may want to add these types of energy at a later time.

3.8 Process Performance Indicators

In the metals industry we need to look at process performance indicators. Knowing how much energy a furnace uses is good but we need to understand the relationship to both mass and process temperature.

3.8.1 Theoretical Energy Requirement

Theoretical energy requirement is the amount of energy required to perform the required work in a perfect world. My old thermodynamics book would give the equation for energy:

$$Q = mC\Delta T \tag{3.18}$$

where

Q = energy
m = mass
C = specific heat
ΔT = Change in temperature

In this case, we want to understand the energy per unit mass (Q/m). This would be in units of kWh/t or MJ/t or Btu/lb. Because we are changing phases, the equation gets more complicated. For every metal and specific alloy of metal, there are certain properties. These include the amount of energy required to heat that metal in the different phases and the melting point of the metal. It takes heat to bring the solid metal from room temperature to the melting point. More heat is required to overcome the latent heat requirement at the phase change. Finally, additional heat is needed to bring the molten metal to the final temperature.

$$Q/m = C_s \times (T_m\, T_0) + L + C_L \times (T_f - T_m) \tag{3.19}$$

where

C_s = specific heat or heat capacity (solid)
C_L = specific heat or heat capacity (liquid)
T_0 = temperature (original or starting)
T_m = temperature (melting)
T_f = temperature (final)
L = latent heat of fusion = energy required to change phases (solid to liquid state)

For example, to bring solid aluminum from room temperature (60°F) to the melting point (1,220°F) would take:

$$Q/m = (T_m - T_0) \times C_S \tag{3.20}$$

$$Q/m = (1{,}220 - 60)°\text{F} \times 0.248 \text{ Btu/lb/F} = 288 \text{ Btu/lb} \tag{3.21}$$

The latent heat needed for the phase transformation for aluminum is 169 Btu/lb. The heat required to bring the molten metal to final bath temperature of 1,350°F (Figure 3.21) is

$$Q/m = (T_f - T_m) \times C_l \tag{3.22}$$

$$Q/m = (1{,}350 - 1{,}220)°\text{F} \times 0.26 \text{ Btu/lb/°F} = 34 \text{ Btu/lb} \tag{3.23}$$

The total heat required to melt and bring aluminum to 1,350°F is:

$$Q/m = 288 + 169 + 34 = 491 \text{ Btu/lb} \tag{3.24}$$

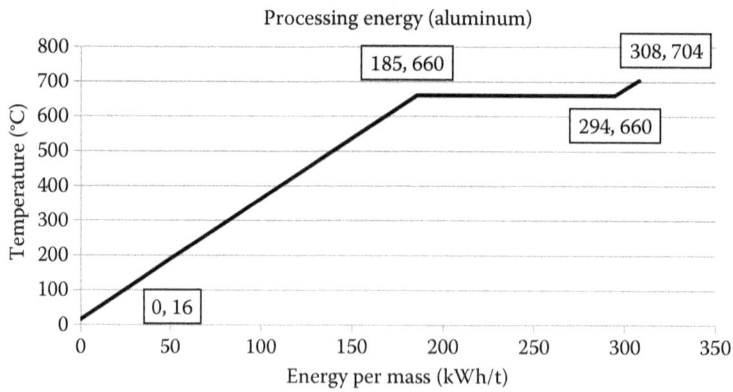

Figure 3.21 Energy used to melt aluminum.

Using metric terms, the example calculation for aluminum would be

$$
\begin{aligned}
Q/m &= [1.04 \text{ kJ/kg/°C} \times (660°\text{C} - 16°\text{C})] \\
&\quad + 392 \text{ kJ/kg} + [1.086 \text{ kJ/kg/C} \times (732°\text{C} - 660°\text{C})] \\
&= 670 \text{ kJ/kg} + 392 \text{ kJ/kg} + 78 \text{ kJ/kg} = 1{,}140 \text{ kJ/kg} \\
&= 1{,}140 \text{ MJ/t} = 317 \text{ kWh/t}
\end{aligned}
\tag{3.25}
$$

If only heat treatment is occurring, calculating the theoretical energy requirement is much easier because we don't have to worry about latent heat or molten specific heat.

$$Q/m = (T_f - T_0) \times C_s \tag{3.26}$$

So, if you are heating carbon steel from room temperature to 1,500°F (816°C):

$$Q/m = (1{,}500 - 60)°\text{F} \times 0.15 \text{ Btu/lb/°F} = 216 \text{ Btu/lb} \tag{3.27}$$

$$Q/m = (816 - 16)°\text{C} \times 0.63 \text{ kJ/kg/°C} = 504 \text{ kJ/kg} = 504 \text{ MJ/t} \tag{3.28}$$

This is the theoretical energy required for the process. To understand the actual efficiency of a process, different measurements are used: energy intensity and energy efficiency.

POTENTIAL PITFALLS!

Specific heat changes over the temperature range, but a single number is used for convenience. Values for these numbers differ per alloy. Values also seem to vary per reference source. Please check!

The data in Table 3.4 is from several nonvalidated sources and are ranges. Research your particular material and alloy for specific numbers. ***Check these numbers before using them!***

Table 3.4 Heat Properties of Metals

Metal	*Specific Heat C_s*		*Melting Point T_m*		*Latent Heat*		*Specific Heat C_L*	
	kJ/kg/°C	*Btu/lb/°F*	*°C*	*°F*	*kJ/kg*	*Btu/lb*	*kJ/kg/°C*	*Btu/lb/°F*
Aluminum	1.04	0.248	660	1,220	392	169	1.086	0.26
Carbon steel	0.63	0.15	1,540	2,800	139	60		
Copper	0.43	0.104	1,083	1,982	211	91	0.464	0.111
Iron (gray cast)	0.79	0.19	1,230	2,246	96	41		
Lead	0.13	0.032	327	621	23	10	0.142	0.034
Magnesium	1.14	0.272	651	1,204	195	84	1.111	0.266
Molybdenum	0.25	0.06	2,621	4,750	292	126		
Nickel	0.56	0.134	1,451	2,644	304	131	0.556	0.133
Platinum	0.13	0.03	1,771	3,220	113	48.6		
Silicon	0.71	0.17	1,411	2,572				
Titanium	0.54	0.13	1,671	3,040				

3.8.2 Energy Intensity

Energy intensity (also called specific energy) is the ratio of the energy used during a heating operation to the product mass that absorbs the energy. The mass is normally in metric tons or kilograms or pounds charged into the process.

$$18 \text{ MMBtu} \times 1,000,000 \text{ Btu/MMBtu} \div 10,000 \text{ lb}$$
$$= 1,800 \text{ Btu/lb energy intensity} \quad (3.29)$$

$$5,275 \text{ kWh} \div 4.535 \text{ t} = 1,163 \text{ kWh/t} \quad (3.30)$$

3.8.3 Energy Efficiency

Energy efficiency (%) is the theoretical energy requirement divided by actual energy usage. It can also be called thermal efficiency, meaning the ratio of the energy absorbed by the material being processed to the total energy entering the system.

$$491 \text{ Btu/lb} \div 1,800 \text{ Btu/lb} = 27\% \text{ energy efficiency} \quad (3.31)$$

$$317 \text{ kWh/t} \div 1,163 \text{ kWh/t} = 27\% \text{ energy efficiency} \quad (3.32)$$

Energy intensity is typically used to track performance within an energy program. Energy efficiency is a good way to understand the difference between heat treat furnaces that heat metal to different temperatures. For instance, the energy efficiency of two heat treat furnaces may be the same even though the energy intensity may be different since one furnace may heat metal to 1,000°F (538°C) while the second may heat to 300°F (149°C).

In general, energy intensity is used for melting furnaces while energy efficiency tends to be used more in heat treating.

POTENTIAL PITFALLS!

You need to define whether *energy intensity is over a single batch/melting cycle or over a month.* Energy intensity is better (lower) when including only melting energy. Over a month there will be holding time and other inefficiencies. Batch energy intensity is good when running specific tests like buy-offs or process changes. Monthly energy intensity is better to understand the whole process for energy management programs.

Keep in mind the theoretical energy requirement as you monitor furnace energy intensity (Figure 3.22). If the performance starts improving, and especially if it reaches *better than theoretical (perfect), there is a metering problem*!

3.9 Key Performance Indicators

Setting the plant Key Performance Indicators (KPIs) is critical to the energy program. Take your time in developing the best measures for your company and your process. Each variable needs to be carefully defined and understood. Different measures may be used for different audiences inside and outside your company.

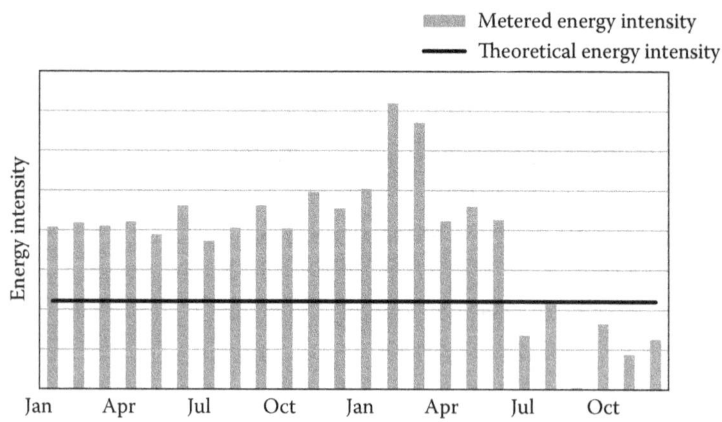

Figure 3.22 Furnace energy intensity.

3.9.1 Units

The energy unit used in reports and for analysis needs to be agreed on so that there is no confusion. This could be in kilowatt-hour (kWh), megajoules (MJ), or British thermal units (Btu). Varied units may be in use within international corporations by the different plants. Different energy types (electricity vs. natural gas vs. coal) tend to use different energy units. For instance, electricity is normally in kWh. Natural gas in the United States is in Dth or MMBtu. Natural gas in Canada or Europe is normally in MJ. The heat content of natural gas in the United States is given in HHV while Europe gas is given in LHV. The unit for mass also needs to be agreed on (pound, kilogram, short ton, long ton, metric ton) so that all plants are using the same unit for energy reporting.

POTENTIAL PITFALL!

Make sure you know if you are using U.S./short ton (2,000 lb) or Imperial/long ton (2,240 lb) or metric ton (2,204.6 lb). Once I had a long discussion with someone from Europe who didn't believe anyone would use any ton measurement other than metric tons. Don't take the term "ton" for granted!

The energy manager may need to convert units after getting local units from the plant. If you ask people at each plant to provide data in a common unit, make sure they know the proper conversion formulas.

Make sure that you check that units are reported for the same time period. Energy bills frequently report energy use starting in the middle of the month (i.e., 15th of the month). The data for the mass and energy need to be for the same time period for the analysis to be correct.

Remember, you don't want to have the problem such as when NASA lost a Mars spacecraft when one team used metric units and the second team used English units. Spend the time you need and clarify units from all your sources!

3.9.2 Plant or Process Performance Indicators

There are different methods of tracking energy performance. Some are simple and some take into account different factors such as production and outside temperature to better understand when the process is out-of-control.

3.9.2.1 Energy Used

This is very basic look at energy used every month (kWh or joules or MMBtu). It measures energy use but doesn't tell you why. Short months such as February will look better than longer months with 31 days. Months with high production will show high energy use. Cold months will look worse. It's hard to tell you if you are getting more efficient (Figure 3.23).

3.9.2.2 Energy per Unit

"Energy per unit" measures efficiency of the process only if product units (castings or ingots) have similar weights. In the metals industry, this is rare (Figure 3.24). The calculation is

$$\text{"Energy per unit"} = \text{Energy} \div \text{Units out} \tag{3.33}$$

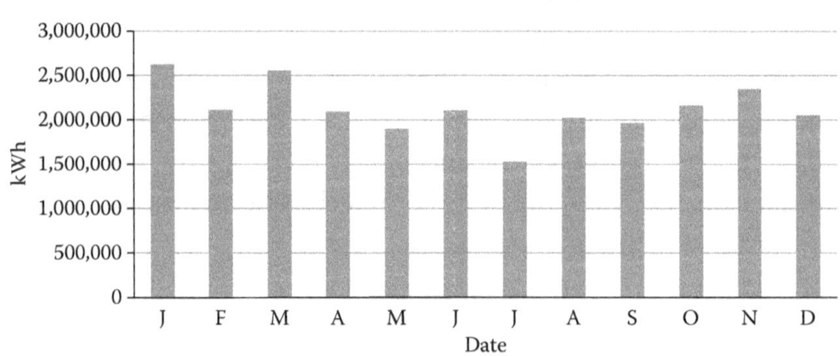

Figure 3.23 Monthly energy usage.

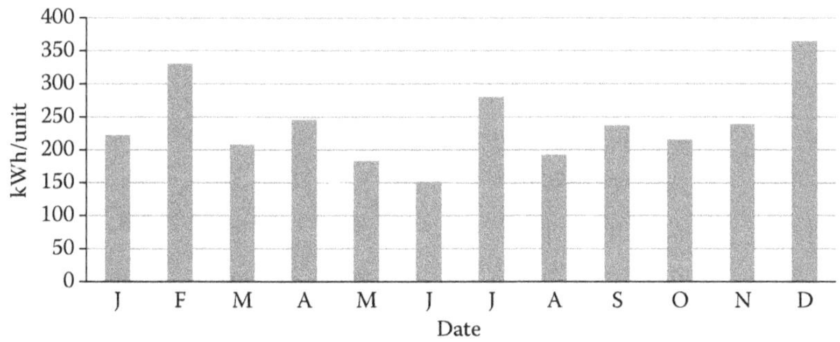

Figure 3.24 Monthly energy per unit.

3.9.2.3 *Energy per Mass Out*

"Energy per mass out" (kWh/t or MJ/t or Btu/lb) measures the efficiency of the process or plant, which is affected by yield, scrap, metal loss, and energy efficiency (Figure 3.25). This is the method used by many outside agencies such as the U.S. Department of Energy (DOE). It truly tells you how efficiently you are producing your final product. This can also be called Energy intensity.

Mass out needs to be well defined. Only good product is normally used, not scrapped material. Packed tons or finished tons are better than shipped tons because Work In Progress (WIP) can skew the analysis. The calculation is

$$\text{"Energy per mass out"} = \text{Energy} \div \text{Mass out} \qquad (3.34)$$

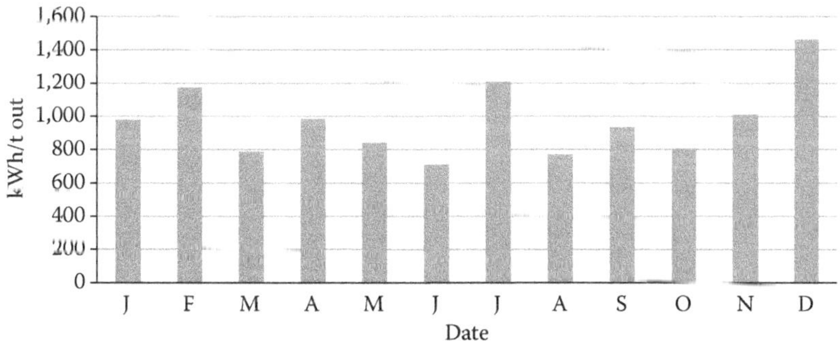

Figure 3.25 Monthly energy per mass out.

3.9.2.4 Energy per Mass In

"Energy per mass in" (kWh/t or MJ/t or Btu/lb) measures energy efficiency of the equipment or process (Figure 3.26). Most energy programs do not work on yield, scrap, or melt loss. It is the best way to understand the efficiency of your energy program for the metals industry. Again, mass in needs to be defined. This is normally charged tons and can include prime material, external scrap, internal scrap, and alloy materials. Molten charge tons are normally not included because little energy is required to process them in the casthouse. If no melting is performed in the plant, this would be metal into the plant for heat treatment or processing.

This is the best method to monitor each separate furnace or line; it is using the material that requires energy to heat. The calculation is:

$$\text{"Energy per mass in"} = \text{Energy} \div \text{Mass in} \tag{3.35}$$

3.9.2.5 Normalized Energy Intensity

The biggest problem with regular energy intensity (kWh/t) is that it varies greatly depending on production levels. Normalized energy intensity is calculated based on previous performance and gives the energy intensity if normal production was made.

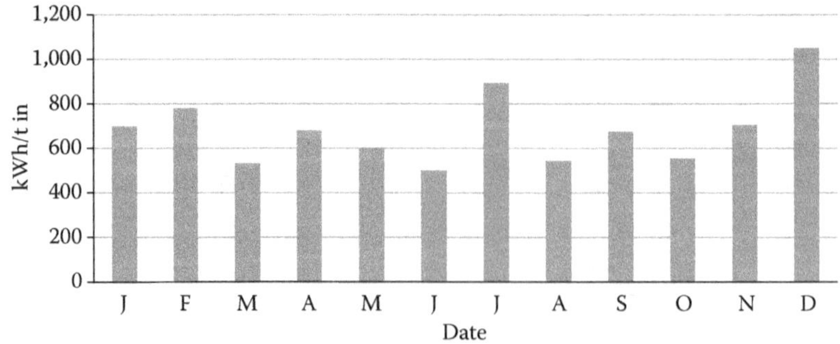

Figure 3.26 Monthly energy per mass in.

To do this, graph and trend the previous years' data showing energy versus production. In Excel®, add a linear trendline and display the equation. The formula will be in the form of $y = mx + b$. In most cases, more energy is used for every ton charged. Use the value of m for the calculations. In this example (Figure 3.27), for every ton another 228.6 kWh is used. Looking at the graph, using a normalized value of 4,000 tons is around average.

To normalize the values in this example, use the following calculation to calculate this value. The monthly numbers are plotted in Figure 3.28.

$$(\text{kWh} + (4{,}000 - \text{tons in}) \times 228.6) \div 4{,}000 \tag{3.36}$$

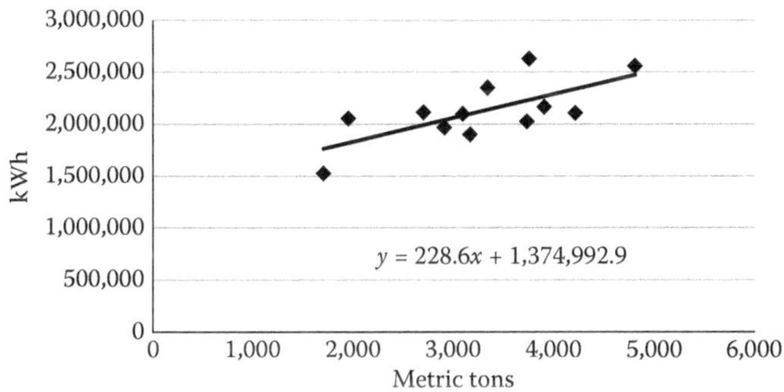

Figure 3.27 Energy versus mass in.

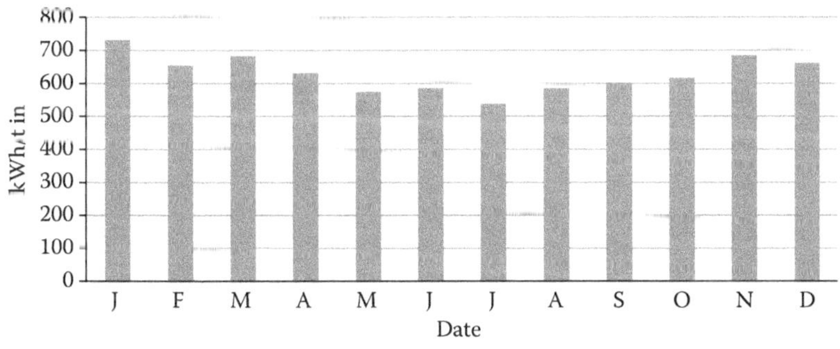

Figure 3.28 Normalized energy intensity.

3.9.2.6 Standard Energy

Standardized values compare expected energy use to actual energy use at that production level. In this case, the higher the number is, the better the efficiency. Standard energy is similar to the previous calculation. Expected gas usage is calculated from the formula of the line on the graph. So, in this example

$$\text{Expected gas usage} = 228.6 \times \text{Tons in} + 1{,}274{,}993 \quad (3.37)$$

Some companies simplify this by using a standard amount of energy per mass. In Excel®, set the intercept to zero for the best fit. This method is not as accurate but it is a simple calculation (Figure 3.29).

$$\text{Expected gas usage} = 621.9 \times \text{Tons in} \quad (3.38)$$

By either method, the expected gas usage is compared to actual gas usage for the month. Figure 3.30 shows the energy standard based on a nonzero intercept.

$$\text{Standard} = \text{Expected gas usage} \div \text{Actual gas usage} \times 100\% \quad (3.39)$$

This method does well in handling changes in production, but it doesn't handle cold weather. Figure 3.30 shows better performance during the summer because no energy is needed for plant heat.

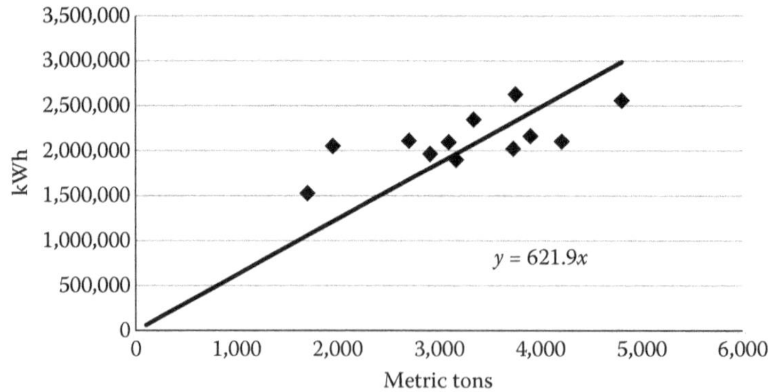

Figure 3.29 Energy versus mass in.

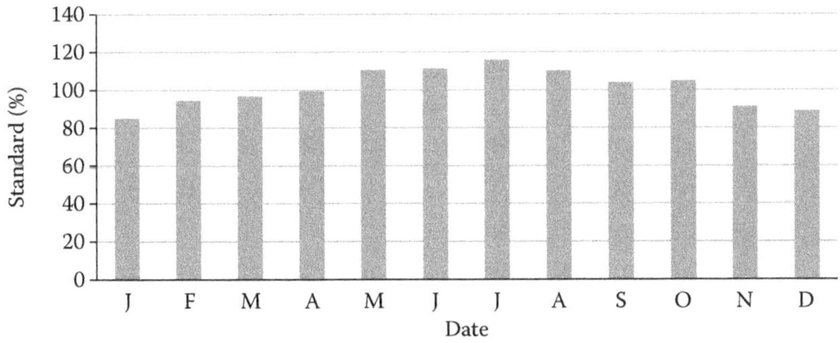

Figure 3.30 Energy standard.

3.9.2.7 Production and Temperature Standard

The biggest issue with using mass out and mass in is if material is purchased at different stages, or if product comes out from different stages of the process or if the product mix changes. For instance, as shown in Figure 3.31, some metal may be melted and cast into billet. The billet may be sold at this point. Or the melting stage may be skipped on some product and it may be purchased from outside. The billet may be processed with extrusion presses and heat treat furnaces. The finished extrusions may be sold at this point or extrusions purchased from outside the plant may enter the process stream. The extrusions may be machined or painted before being sold. Each stage of this process has different energy

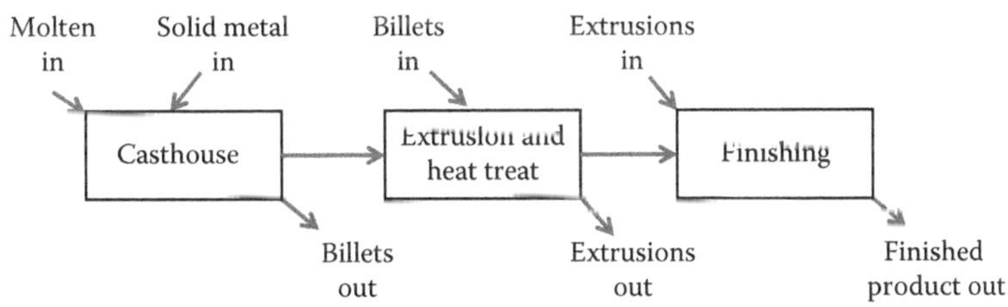

Figure 3.31 Metal in and out industrial processes.

requirements. "Mass out" and "mass in" are two simplified numbers and may not be the best method of tracking plant performance.

A production and temperature standard (Figure 3.32) can be calculated using mass into each major process (melting, heat treating, machining, etc.). Outside temperature can be used to calculate the expected energy usage for plant heat. This method shows changes without being affected by the mix. It is the most complicated method but gives the truest results to know when problems are occurring or improvements have been made. If you have a simple process or the mix doesn't change, this method may be more complicated than you need. However, it works for many plants in the metals industry with varied products.

Multiple indicators may be used to understand the process better. Many times it depends on the particular plant. For instance, if the plant is in a temperate zone, outside temperature is not critical. If yield varies greatly, input pounds may be more important. If a product mix changes so that more metal is heat treated and less is melted in a specific month, a more complex standard calculating the expected gas usage depending on mass and process is appropriate.

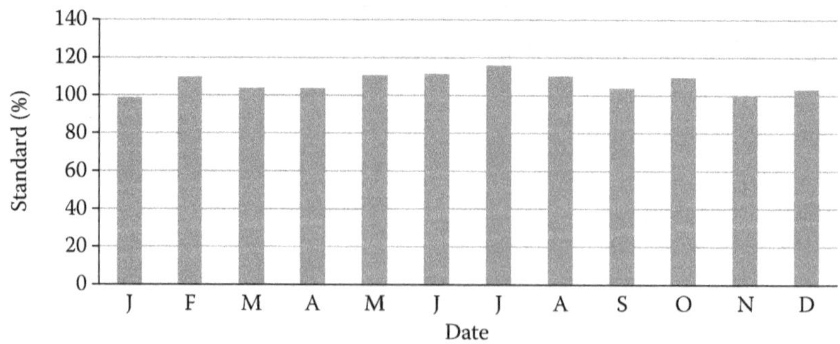

Figure 3.32 Temperature energy standard.

3.9.3 *Energy Costs*

3.9.3.1 Purchased Cost

Knowing the current and forecasted cost of energy is important to determine return on investment (ROI) on energy projects. Given the volatility of energy prices, it's good to recheck numbers at least every six months. Another number you may need is the accounting cost of energy. This may have been set years ago. Some companies use the accounting cost for project cost justification.

3.9.3.2 Cost per Mass

Cost per mass (€/t or $/lb) is very important to most plant managers. This is the cost to process each ton through the plant. However, it is best to keep costs out of the energy KPIs. A plant can control energy usage, but not the cost of energy.

3.9.3.3 Energy Comparison

It is good to understand energy costs of alternative fuels such as landfill gas or fuel oils. Back in the 1970s and 1980s, energy costs peaked in the United States. Everyone went out and bought dual-fuel burners so the fuel source could be changed from fuel oils to natural gas depending on current prices. Some of the burners out there burn liquid fuel, natural gas, or pulverized coal. Changing energy sources is not as frequent now in the United States, but it happens more outside the country and may happen again in the United States.

Sometimes changing your fuel can be a good option such as using locally produced landfill gas. With low natural gas prices in the United States, landfill gas is now less cost effective. Depending on your location and tax rebates, wind or solar electricity may make sense. Combined heat and power

(CHP) units may also make sense for your plant. However, remember that our business is to make metal, not electricity. Most plant maintenance departments are good at maintaining furnaces but working on a wind turbine is outside their expertise.

Remember, Table 3.5 is just a snapshot in time and location. Understand *your* costs and keep up to date on the options!

Understanding the implications and greenhouse gases formed (tons or lbs or metric tons CO_2) is important too. Using electricity instead of a fossil fuel eliminates CO_2 formation at your plant but moves it to the power plant. Owing to inefficiencies and losses, the amount of CO_2 formed at the power plant is normally two to three times greater than if the fossil fuel is used directly at your plant. However, the power plant fuel makes a major difference. If the power plant uses wind, hydropower, nuclear, or solar, the direct formation of CO_2 disappears. Table 3.6 shows the different rates of CO_2 formation. Of these fuels, burning natural gas at your plant currently generates the least greenhouse gases into the environment.

Table 3.5 Example Energy Cost Comparisons

Energy Source	*Cost*	*Example Heat Content*	*Cost ($/kWh)*
Electricity	$0.0689/kWh	N/A	$0.069
Natural gas	$2.28/MMbtu	N/A	$0.008
Landfill gas	$2.50/MMbtu	N/A	$0.009
Crude oil	$39.85/barrel	5,800,000 Btu/barrel	$0.023
Gasoline	$1.31/gallon	123,000 Btu/gallon	$0.036
Heating oil	$0.97/gallon	139,000 Btu/gallon	$0.024
Diesel	$1.15/gallon	139,000 Btu/gallon	$0.028
Propane	$0.46/gallon	91,330 Btu/gallon	$0.017
Coal	$42.25/short ton	12,500 Btu/lb	$0.006

Source: www.eia.gov (accessed March 15, 2016).

Table 3.6 CO_2 Formation

Fuel	*Power Plant Fuel*	*CO_2 (lb/MMBtu)*	*CO_2 (lb/kWh)*	*CO_2 (mt/TJ)*
Electricity	Coal (bituminous)		2.07	0.26
	Natural gas		1.22	0.15
	Residual oil (No. 6)		1.76	0.22
Coal (bituminous)		205.7	0.70	0.09
Coal (anthracite)		228.6	0.78	0.10
Natural gas		117.0	0.40	0.05
Diesel fuel		161.3	0.55	0.07

Source: http://www.eia.gov/tools/faqs/ (accessed June 30, 2016).

POTENTIAL PITFALL!

In general, electrical furnaces are more energy efficient than natural gas furnaces. However, many times the cost of electricity outweighs the difference. Make sure energy evaluations on new furnaces include the projected cost of energy.

3.10 Setting Overall Goals

3.10.1 Why Do We Worry about Goals?

Setting goals

1. Sets the tone for improvement
2. Communicates intended results
3. Helps identify progress and setbacks
4. Fosters ownership of energy management, and motivates staff support and action
5. Creates a schedule for maintenance and upgrade activities
6. Measures the success of an energy management program

3.10.2 Keys to Goal Setting

1. Reasonable goals are required! Don't ask your plant to reach goals they can't reach within the time period and capital resources.
2. Determine scope and definition: Energy use, energy intensity using input mass, energy intensity using output mass, and so forth.
3. Annual improvement can be 2%–5%. A 2% compounded average is used for many companies.
4. Consider long-term goals (e.g., 10% over 5 years). This gives management more time and freedom to reach the goal.
5. Identify organizational and time parameters for goals.
6. Estimate potential for improvement.
7. Consider threshold goals or stretch goals—3% per year?
8. Create performance targets for each facility, department, and major process.

Table 3.7 shows published energy goals from some larger U.S. companies. An improvement of 2% per year sounds small but quickly adds up.

3.11 Benchmarks

Just knowing the gas usage, energy intensity, or energy efficiency doesn't tell you if you are running well. Comparisons need to be made. Behavioral science has shown that knowing how others are performing makes a change in our behavior. Knowing benchmarks for furnaces, processes, and buildings will make a difference!

POTENTIAL PITFALL!

A plant installed the latest furnaces and equipment for heat treating their product. Meters were installed but not read. After

Table 3.7 Company Goals

3M	20% reduction in energy use per unit product for 2005–2010; 19.8% energy reduction per unit achieved through 2009; efficiency calculated by dividing 3M net sales by 3M total energy use
BOEING	25% energy intensity improvement by 2012 versus 2007; 5.3% energy intensity reduction by 2009
Ford	15% energy reduction corporate-wide by 2011 versus 2006; 13.6% energy reduction through 2009, 38% since 2000
MERCK	25% energy intensity reduction by 2009 versus 2004; 30% reduction achieved through 2009, exceeding goals; intensity measured in million BTU/square foot
PEPSICO	20% electricity, 25% fuel consumption reduction for 2006–2015; 5.1% electricity reduction in 2009, 36% since 1999
PPG	2.5% energy reduction per year for 10 years versus 2006; in terms of total net purchased energy per ton of saleable product

an energy program was implemented, the plant began measuring energy use, calculated energy intensity, and benchmarked to other heat treat furnaces in the company. Instead of being the best, they were found to have the least efficient heat treat furnaces in the corporation (last four furnaces in Figure 3.33).

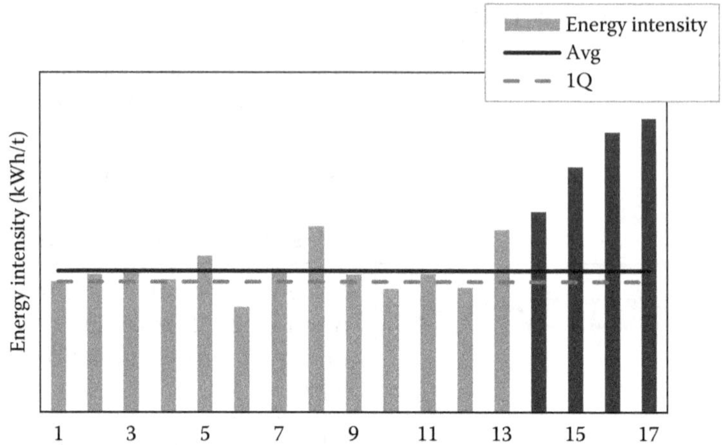

Figure 3.33 Energy intensity for heat treat furnaces.

Further work identified issues with the furnaces and process. Don't rely on suppliers, old data, or "feelings." Measure and compare!

Compare performance against

Historical. Past performance of the particular furnace or process.

Within plant or between plants. Look at similar processes. The first quartile (1Q) is a good benchmark. This means that 25% of the data is within the first quartile (1Q). This number is a challenge, but possible! A simple way of calculating the first quartile is to use to the Excel® function—quartile(array,1).

Figures 3.34 and 3.35 show a histogram and a bar graph of energy intensity for specific types of melting furnaces. The average is shown along with a benchmark at the first quartile.

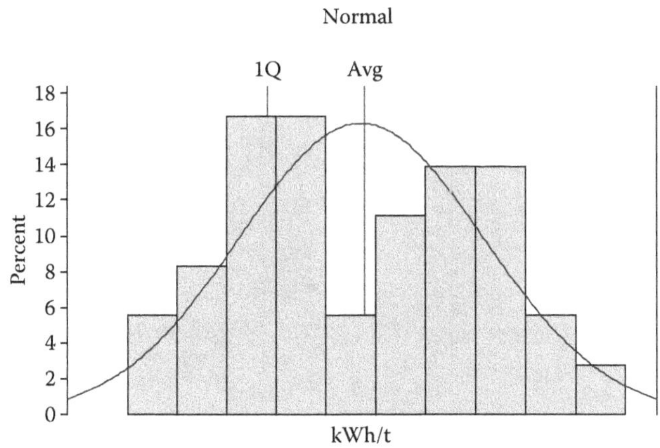

Figure 3.34 Histogram: Energy intensity average and 1Q for furnace type A.

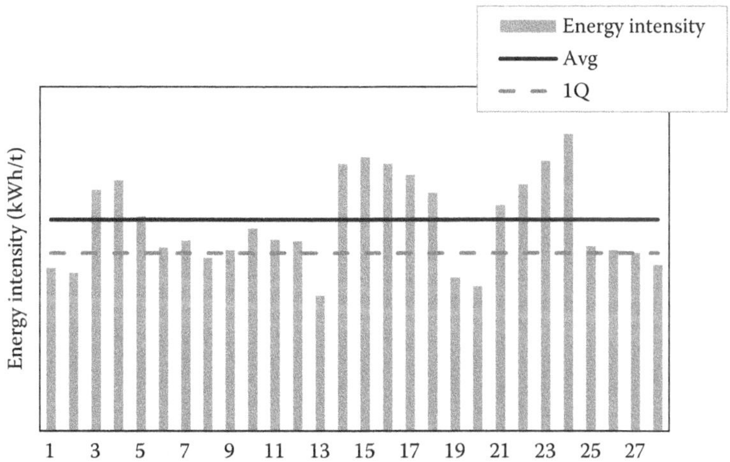

Figure 3.35 Bar graph: Energy intensity average and 1Q for furnace type A.

The best 25% of the furnaces lie below this line. Theoretically, all the furnaces could reach this level.

Figures 3.36 and 3.37 show a histogram and a bar graph of energy efficiency plotted for heat treat furnaces. These furnaces heat the metal to different temperatures so a comparison

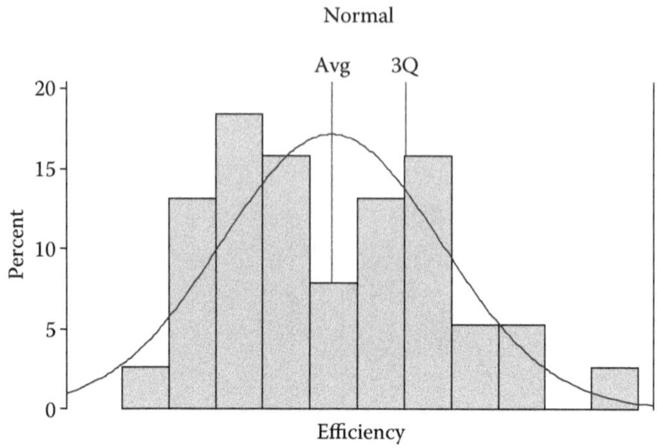

Figure 3.36 Histogram: Energy efficiency average and 3Q for furnace type B.

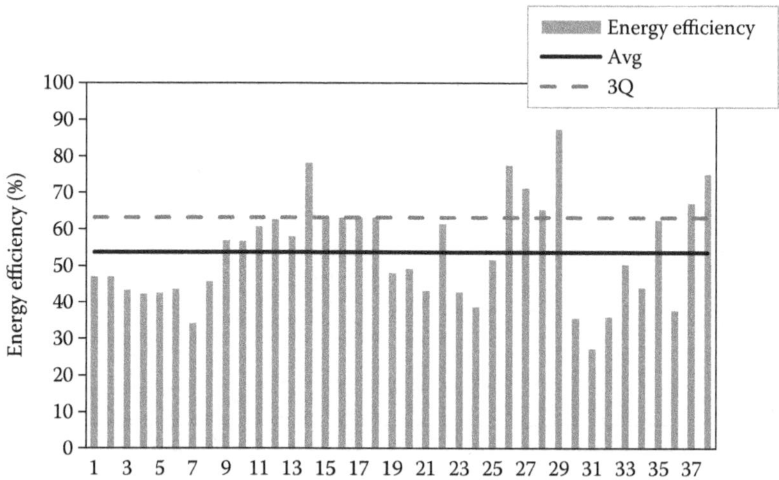

Figure 3.37 Bar graph: Energy efficiency average and 3Q for furnace type B.

of energy intensity (kWh/t) would be inappropriate. Energy efficiency (%) is based on the energy required to bring the metal up to the final temperature. The average is shown along with the benchmark. In this case, the benchmark is set at the third quartile because higher efficiencies are the better furnaces.

Externally to Industry Standards and Best in Class talk to vendors and read papers. Some possible resources are listed below.

1. Aluminum: *NADCA Energy Saving Manual.* Order through www.diecasting.org.
2. Aluminum: "Bandwidth Study on Energy Use and Potential Energy Savings Opportunities in the Manufacturing of Lightweight Materials: Aluminum," Energetics Incorporated, DOE Advanced Manufacturing Office, Draft March 2016.
3. Aluminum: C. Belt (2015), "Current State of Aluminum Melting and Holding Furnaces in Industry," *JOM*, 67(11): 2690–2695.
4. Iron and steel: "Bandwidth Study on Energy Use and Potential Energy Saving Opportunities in U.S. Iron and Steel Manufacturing," Energetics Incorporated, DOE Advanced Manufacturing Office, June 2015.
5. Iron and steel: E. Worrell et al., "Energy Efficiency Improvement and Cost Saving Opportunities for the U.S. Iron and Steel Industry," U.S. Environmental Protection Agency, 2010.
6. Casting furnaces: J. F. Schifo and J. T. Radia, "Theoretical/ Best Practice Energy Use in Metal Casting Operations," http://www1.eere.energy.gov/manufacturing/resources /metalcasting/pdfs/doebestpractice_052804.PDF.
7. Benchmarking analysis for buildings (ENERGY STAR 2016c)—used for energy and water consumption. Energy performance is ranked on a scale of 1–100 compared to similar buildings nationwide. http://www.energystar.gov /index.cfm?c=evaluate_performance.bus_portfoliomanager.

POTENTIAL PITFALL!

When benchmarking, watch the following:

- Comparing energy intensity for mass in versus mass out can make a substantial difference.

- Fuel type makes a difference in benchmarking. Heat treat furnaces can be natural gas, electric, or a mixture of both types. Sometimes it makes sense to compare cost per ton versus kWh per ton.
- Understand which fuels are included. Often electricity for items such as fans and blowers is ignored when calculating energy intensity. Sometimes it is a small component and is not critical. Other times the hidden use of electricity makes a difference.
- When comparing numbers between the United States and Europe, try to understand if HHV or LHV is being used.
- When comparing departments, make sure you know what processes are included. Does a rolling department include homogenizing, reheat, or anneal? Is the energy used to create nitrogen or other atmosphere gases included in anneal furnaces? Scope is important.

Chapter 4

Analyze Data and Identify Projects

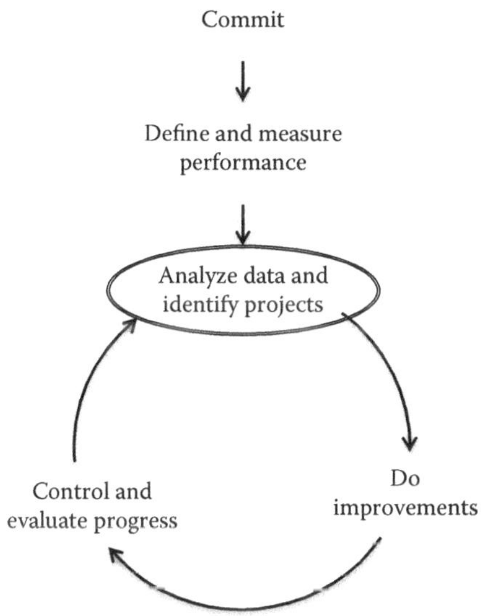

Now that you have all this data, what do you do with it? We're going to discuss data analysis techniques. Energy intensity needs to be understood for processes in the materials industry. Then we're going to start our project list and discuss types of projects. In the materials industry, process heating, waste heat recovery, and furnace testing are critical. Finally, we will discuss technical system analysis. We will discuss many ideas for energy projects throughout this chapter.

4.1 Data Analysis

Before you start identifying projects, you need to *analyze data* to understand which areas to address and what can be done. What is the "secret" of energy savings? There is no secret. Use data, data, and more data! Look at the data multiple ways—by month, day, shift, alloy, gauge, and so on. Identify energy use trends. Look for peaks and valleys. Look for changes over time. Some examples are shown in Figures 4.1 through 4.3.

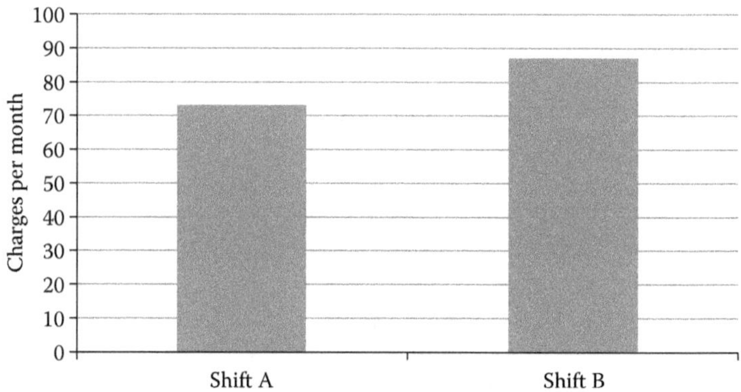

Figure 4.1 Charges per month.

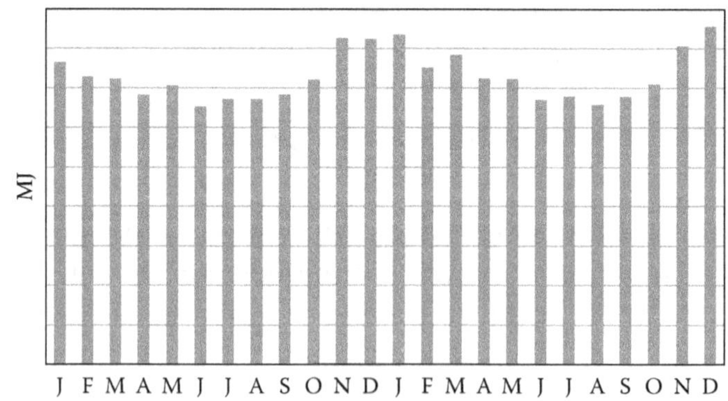

Figure 4.2 Plant gas usage.

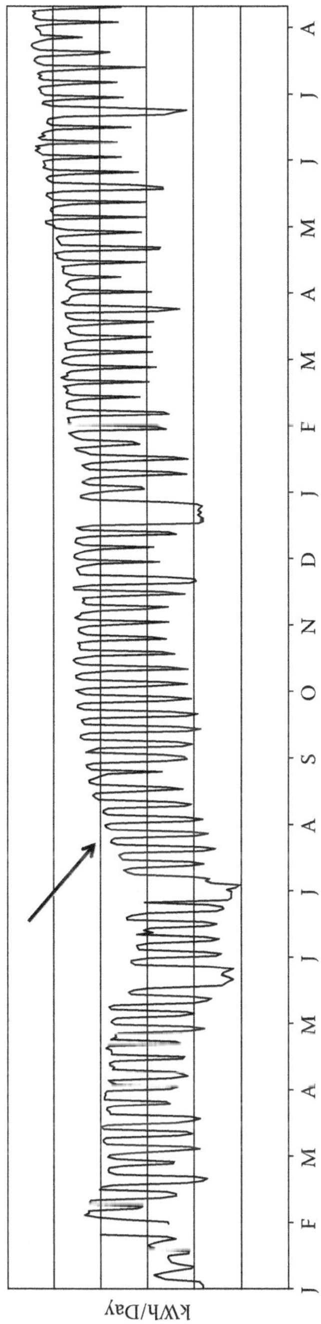

Figure 4.3 Daily electricity.

Table 4.1 Energy Use in the Metals Industry

Energy Use (2006)	*Iron and Steel*		*Alumina and Aluminum*		*Foundries*	
Process	*TBTUs*	*%*	*TBTUs*	*%*	*TBTUs*	*%*
Process heating	723	74.9	101	38.1	106	67.9
Process cooling	1	0.1	3	1.1	1	0.6
Other processes	28	2.9	4	1.5	2	1.3
Electrochemical	9	0.9	120	45.3	2	1.3
Machine drive—pumps	10	1.0	2	0.8	2	1.3
Machine drive—fans	17	1.8	3	1.1	3	1.9
Machine drive—compressed air	18	1.9	3	1.1	3	1.9
Machine drive—materials handling	63	6.5	10	3.8	9	5.8
Machine drive—materials processing	14	1.5	3	1.1	2	1.3
Machine drive—other systems	2	0.2	0	0.0	0	0.0
Facility HVAC	67	6.9	9	3.4	20	12.8
Facility lighting	8	0.8	3	1.1	3	1.9
Other facility support	2	0.2	2	0.8	2	1.3
Onsite transportation	3	0.3	1	0.4	1	0.6
Other nonprocess	0	0.0	1	0.4	0	0.0
Total (TBTUs)	965		265		156	

Source: http://www1.eere.energy.gov/manufacturing/resources/footprints.html (accessed in 2013; link no longer available).

The number of charges into the furnace (Figure 4.1) varied between shifts. More charges mean higher throughput and better energy efficiency. Is there something we can learn from shift B?

Figure 4.2 shows the trend for increased gas usage during the winter months. Reducing energy used to heat the plant may be a viable project.

Figure 4.3 shows daily electrical energy use over time. Electricity use during the week has been increasing since August. Is that due to increased production? Weekend electricity use has also increased (the valleys of the data). Can we improve idling practices on down-days?

You need to identify and rank areas of high energy use. Collect as much data as possible from meters and sort into areas to study. Spend most of your time analyzing your high-energy users. Table 4.1 shows 2006 energy use in different U.S. metal industries by process. It may be very different in your plant!

4.2 Energy Intensity

Many companies consider only energy intensity (kWh/t or Btu/lb) when looking at furnace performance. A more in-depth look is needed. Three different analytical methods can be used, depending on the process (Belt and Peterson 2008): Fixed Energy Usage per Time Unit, Fixed Energy Usage per Production Unit, and Variable Energy Usage per Production Unit. Energy intensity can require special techniques and understanding when performing statistical analysis.

4.2.1 Fixed Energy Usage per Time Unit

These are processes in which the energy used is controlled by time.

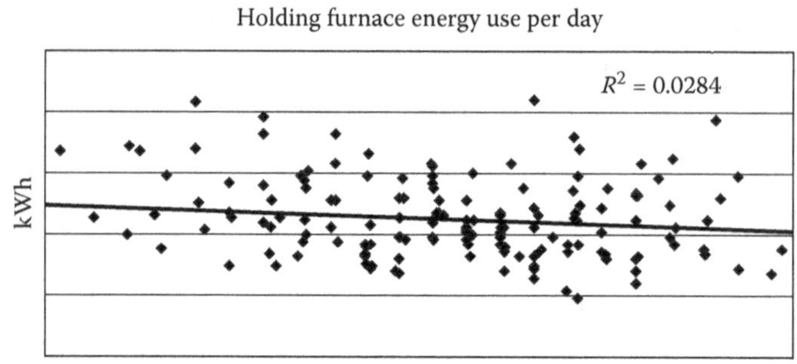

Figure 4.4 Energy use versus production—holding furnace. (Adapted from C. Belt and R. Peterson, "Analysis of Energy Efficiency for Industrial Processes," *EPD Congress 2008*: 53–60. Copyright 2008 by The Minerals, Metals & Materials Society. Used with permission.)

- Examples include plant heat, lighting, and holding furnaces.
- The main purpose of a holding furnace is to replace heat losses. The losses are generally fixed and do not depend on throughput. When a graph is plotted for energy use per day versus throughput (Figure 4.4), there is little to no correlation. In this example, energy use actually slightly decreased with increased throughput.
- For these processes, kWh/t (Btu/lb) is not as effective as kWh/month (Btu/month).
- If production decreases, review equipment of this type. Maybe you can't shut down a holding furnace, but equipment such as lights and fans can be turned off on down days to reduce energy usage.

4.2.2 Fixed Energy Usage per Production Unit

Most processes in the metal industry increase energy use with higher throughput. For every additional ton of product, a roughly fixed amount of energy is used. This is the energy intensity (kWh/t or Btu/lb).

- Examples include annealing and aging furnaces without refractory walls.

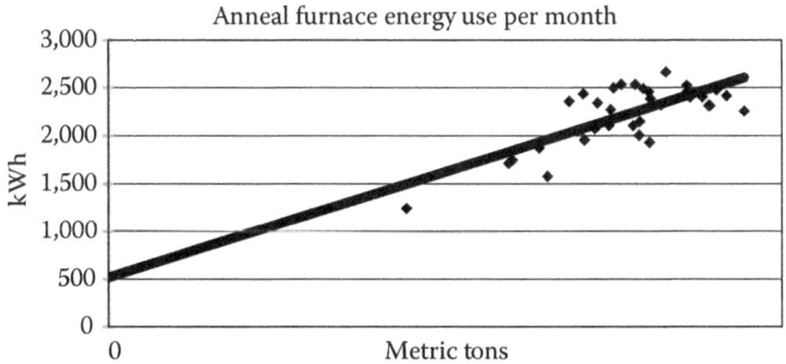

Figure 4.5 Energy use versus production—anneal furnace. (Adapted from C. Belt and R. Peterson, "Analysis of Energy Efficiency for Industrial Processes," *EPD Congress 2008*: 53–60. Copyright 2008 by The Minerals, Metals & Materials Society. Used with permission.)

- In these processes, more production requires higher energy use. When a graph of energy use versus throughput is plotted, a strong correlation is seen (Figure 4.5). Each data point shows the energy and throughput for a single month through this anneal furnace.
- For these processes, the line crosses zero (or close to it). When production is zero, energy use is low. These processes can be idled easily. The process can be shut down and little energy is required to heat up the system.
- If energy intensity is plotted (Figure 4.6), a near-horizontal line is seen. More production doesn't make a major difference in energy intensity (kWh/t or Btu/lb).

4.2.3 Variable Energy Usage per Production Unit

Many processes cannot be idled easily. Holding time occurs due to alloying, charging, maintenance, scheduling issues, and so on. Energy is needed even when holding.

- Examples include many furnaces with refractory such as melting furnaces.

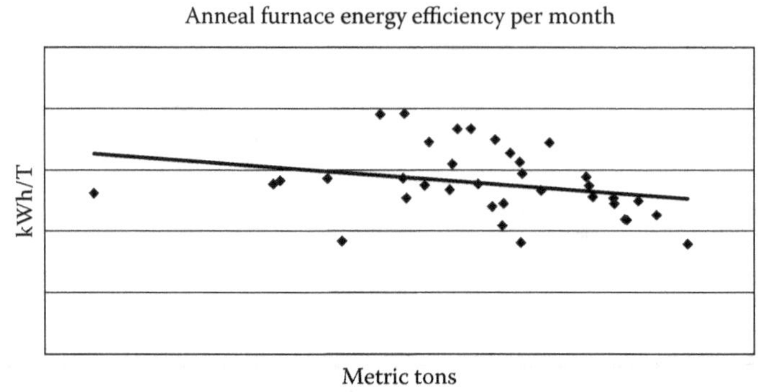

Figure 4.6 Energy intensity versus production—anneal furnace. (Adapted from C. Belt and R. Peterson, "Analysis of Energy Efficiency for Industrial Processes," *EPD Congress 2008*: 53–60. Copyright 2008 by The Minerals, Metals & Materials Society. Used with permission.)

- For these processes, energy use and production per day are correlated, but the line does not cross zero (Figure 4.7). Energy is needed to hold.
- When energy intensity is plotted versus throughput (Figure 4.8), a strong correlation is seen versus throughput. However, the relationship is not linear. Energy intensity is best when throughput is high. At low utilization, energy intensity can be extremely poor!

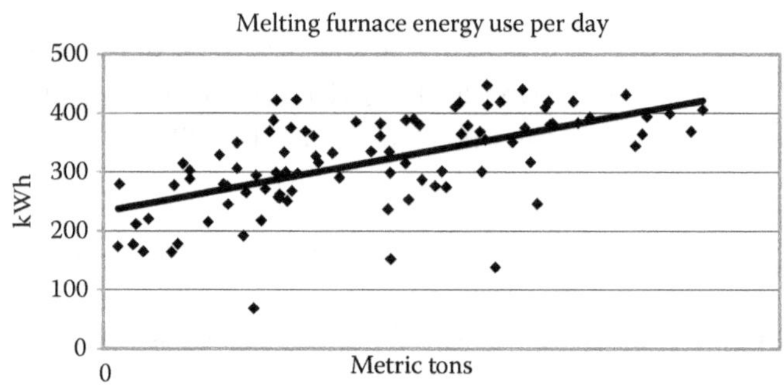

Figure 4.7 Energy use versus production—melting furnace. (Adapted from C. Belt and R. Peterson, "Analysis of Energy Efficiency for Industrial Processes," *EPD Congress 2008*: 53–60. Copyright 2008 by The Minerals, Metals & Materials Society. Used with permission.)

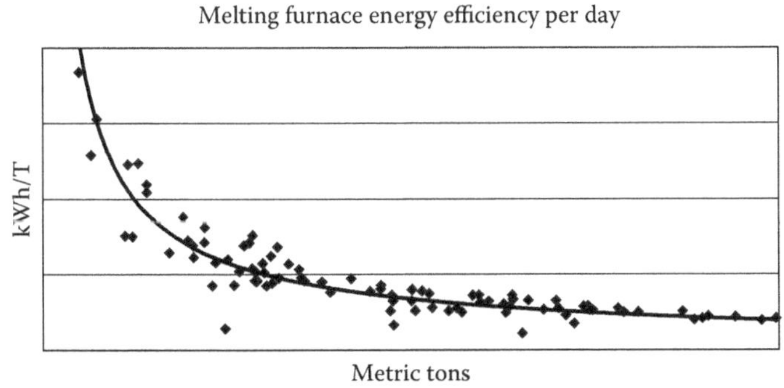

Figure 4.8 Energy intensity versus production—melting furnace. (Adapted from C. Belt and R. Peterson, "Analysis of Energy Efficiency for Industrial Processes," *EPD Congress 2008*: 53–60. Copyright 2012 by The Minerals, Metals & Materials Society. Used with permission.)

Note: When plotting energy intensity versus production in Excel®, use a power curve. In most cases, a linear curve does not fit.

For these processes, it is important to keep production levels high. Anything that reduces throughput (long charge times, bottlenecks in the next process, etc.) will worsen energy intensity. If too many furnaces are running, it may be best to shut down furnaces to optimize the system. It may be better to have four smaller furnaces than two big furnaces, so extra furnaces can be shut down versus running a big furnace at low utilization.

When evaluating new equipment or changes in the process, this relationship must be kept in mind. Because production is difficult to control, it is best to plot energy use versus throughput. Plot energy used versus mass charged data in a program such as Excel® and add a linear line. The goal is to move the line down. Moving the line down means less energy was used for the same throughput. Figure 4.9 shows the results of two trials along with a baseline. In each trial, the trend line was lower, showing that less energy was used for the same production quantity.

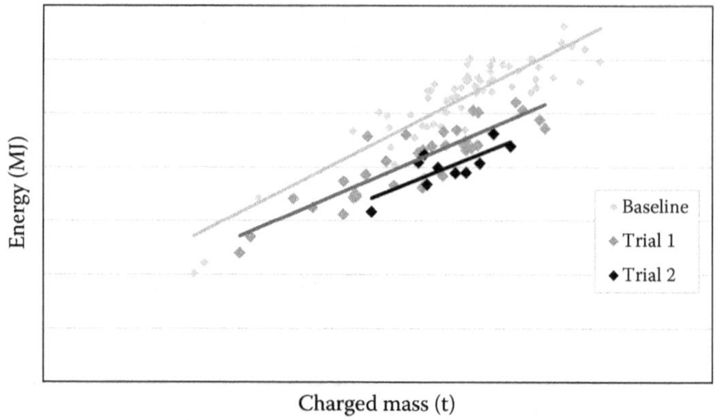

Figure 4.9 Furnace energy trials.

Note that these plots are a good way to explain the performance of trials to management. A visual representation like this can be helpful.

Given the relationship of energy intensity and throughput, there are three ways to improve energy intensity for these processes: high throughput, lowering the intercept (lower heat loss or other energy conservation techniques), or reducing the slope (energy efficiency techniques). In the example shown in Figure 4.10, the improvement reduced the slope. Energy used at low throughput did not change or was slightly worse, but the change made a large difference during the melting process, as shown in high-throughput weeks.

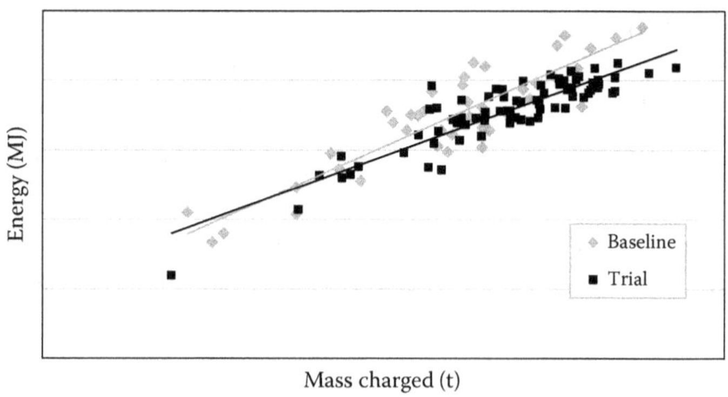

Figure 4.10 Furnace energy trial—high throughput.

POTENTIAL PITFALL!

Here is one method salesmen use to scam you to buy new equipment such as magnets on the gas line. Tests are run with high production but compared to average production data. Make sure you have a plot of current performance and compare trials to this line. Another method is to test with a controlled amount charged both with and without the new equipment.

4.2.4 Statistics

Energy intensity can be difficult to handle with statistics because of the relationship between throughput and energy intensity. Figure 4.11 shows the scatterplot of energy intensity versus mass charged for one furnace. As you can see, the amount of material charged each day varied greatly. This resulted in a large variation in energy intensity.

Figure 4.12 shows the statistics calculated from this data set. A *p*-value of <0.05 for the Anderson–Darling Normality Test means the data set is not normal. You can see the high

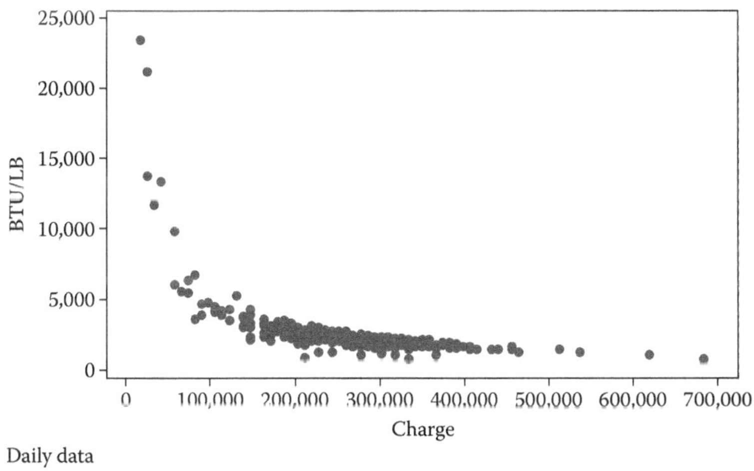

Figure 4.11 Scatterplot of energy intensity versus charged mass.

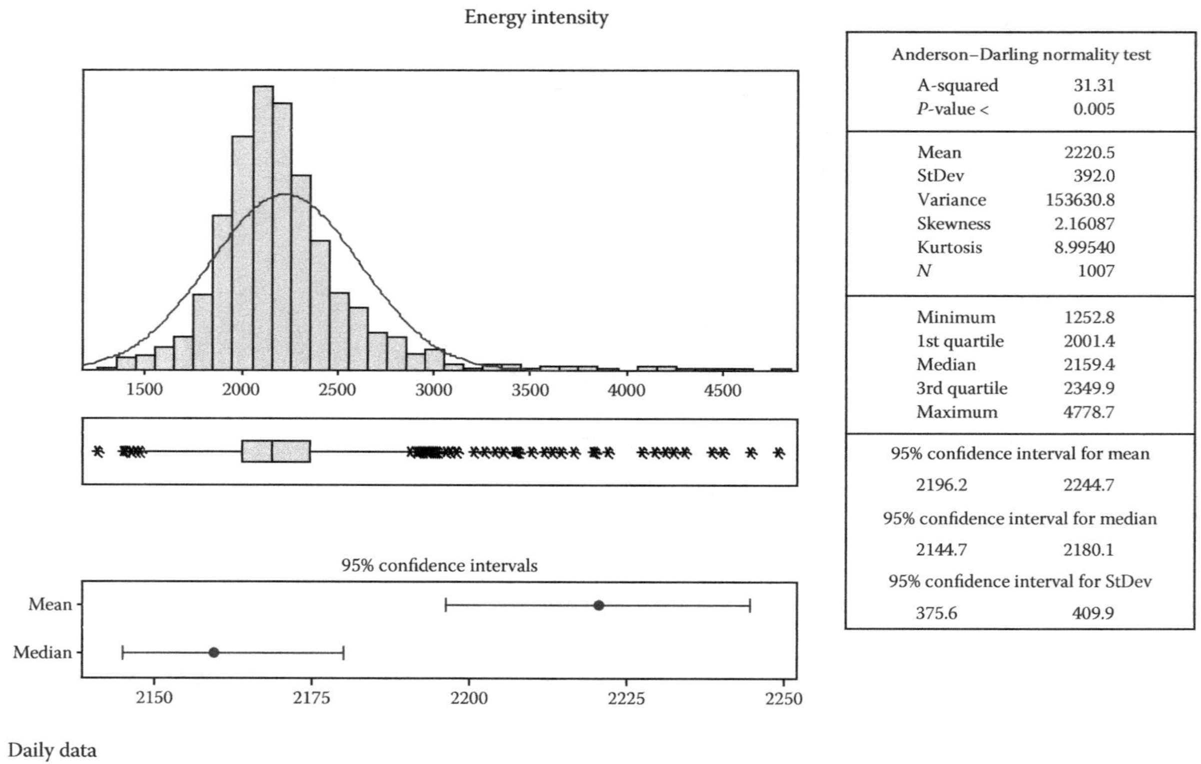

Figure 4.12 Statistical summary of energy intensity—nonnormal data.

amount of skewness with a large tail to the right. These data points are due to days with low production. Making a comparison between two sets of data like this means more data than usual needs to be collected and statistical tests for non-normal data will need to be run. You can also see that there is a difference between the mean and median with this type of data.

Different furnaces and different process require different kinds of statistical analysis. Figure 4.13 shows data from another furnace where the energy intensity is normal ($p > 0.05$) and the mean and medians are almost identical. These data results used monthly data so that the range in throughput was not as high as the previous example. This may also happen on batch data if the batch size falls within a small range. Check your data when you are performing a statistical analysis.

Reporting the average values based on averaging energy intensity (kWh/t) results in an incorrect conclusion. You need to go back and sum energy (kWh) and mass (t) and then recalculate energy intensity (kWh/t) to get the correct average value (see Table 4.2).

Table 4.2 Energy Intensity Calculations

kWh	*Tons*	*kWh/t*
12,000	20	600
14,000	25	560
16,000	30	533
10,000	12	833
15,000	30	500
	Incorrect Avg kWh/t	605
Sum kWh	*Sum t*	*Calc Avg kWh/t*
67,000	117	573

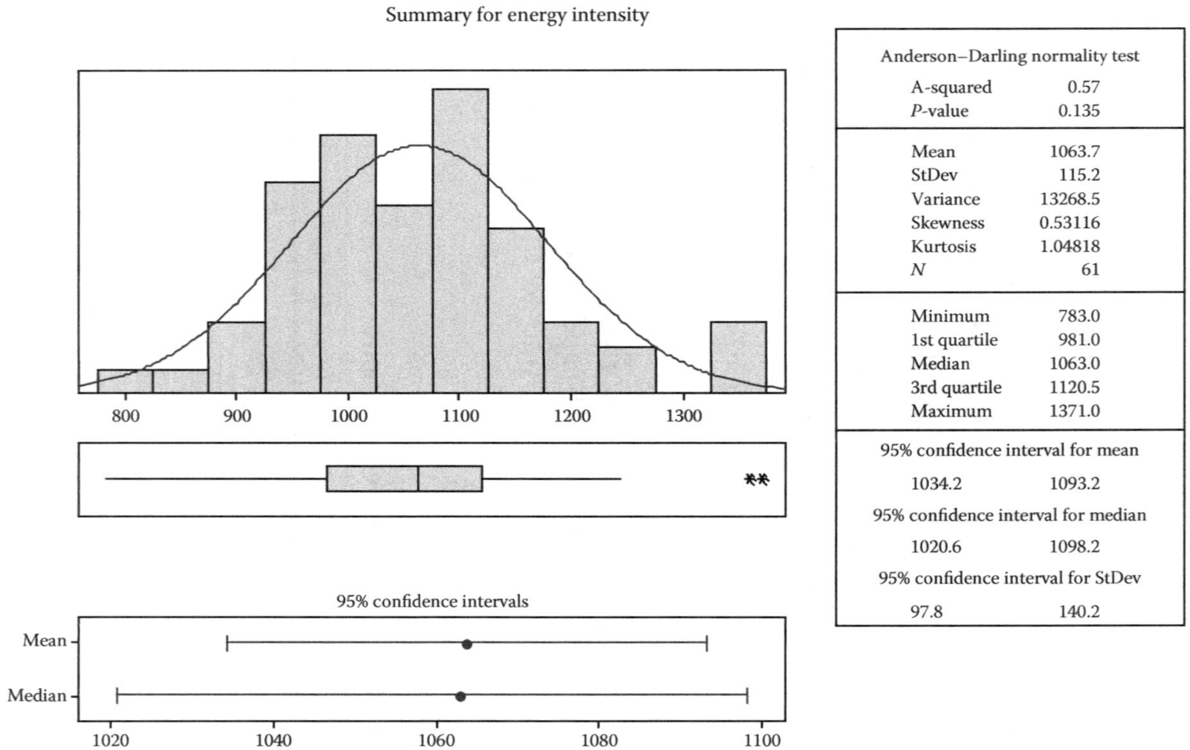

Figure 4.13 Statistical summary of energy intensity—normal data.

4.3 Project List

Once you've analyzed data it is time to consider projects to improve energy efficiency. In general, you want to work on areas that are less energy efficient and have the best chance for cost savings.

Review goals and benchmarks and compare similar types of equipment. Figure 4.14 is a graph of energy intensity for the same type of furnace. Three furnaces really stand out as needing help!

Brainstorm and list opportunities. Don't worry if the ideas are good or bad, just list them.

1. Look at the data!
2. Conduct a facility walk-through. Observe use and note obvious problems. Are furnace doors open unnecessarily? Are lights on in areas that are unused? Is equipment well maintained?
3. Talk to the people. Do they know what the energy use is in their area? Do they have any ideas of how it is being wasted?

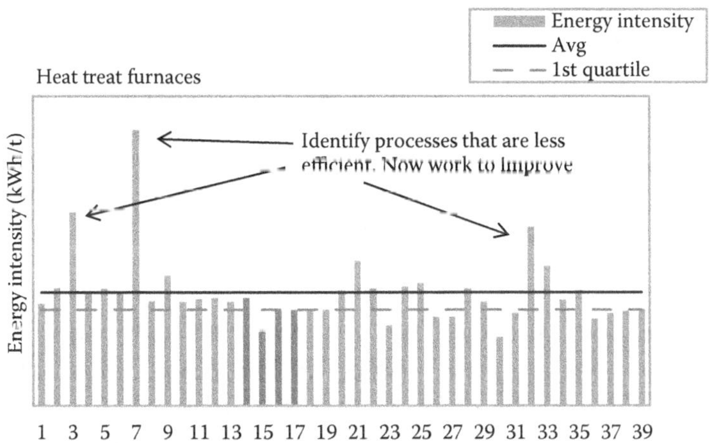

Figure 4.14 Heat treat furnace energy intensity.

4. Watch out for "Just in case" scenarios such as "We keep this hot just in case."
5. Ask why, why, why… ? A Six-sigma tool is called 5-Whys. Keep asking why until you have the root cause or causes. For instance, the air/fuel ratio is off because there is no written preventative maintenance (PM) and no training and no equipment to read flue O_2.
6. Evaluate past projects and best practices.
7. Review technical assessments and audits.

POTENTIAL PITFALLS!

Don't ignore small but multiple gas consumers or users. For instance, crucible/ladle heaters are small but some plants have many of them. One plant kept a spare crucible on heat at all times "just in case." In five years, the spare had never been used. Turning off this burner saved $50,000 per year.

Spend most of your time on your highest gas users but keep in mind the benchmark values for your smaller gas users. Many times the smaller users are ignored and are very inefficient. These projects may be easier to justify in terms of cost.

4.4 Types of Projects

Now, let's consider types of energy projects along with some examples. The first type is the popular no-cost or low-cost projects. This includes projects that address energy price, behavior, simple adjustments, and maintenance. The next type is capital and expense projects. This can include proven technology, trials, and major capital projects. Another type is projects that can be run at the corporate level by the energy manager. The next type is process projects, which include yield improvements, energy value stream mapping, and other methods. A final type of project are ones that may not be cost effective but are done for public relations.

4.4.1 No-Cost or Low-Cost Projects

These are the projects that management loves (the "low-hanging fruit"). They don't cost much, but may take some time. These should be considered first. Some higher cost projects are also listed here as more automated or technical versions. These automated projects will most likely save more money and be more consistent, but implementing the low-cost version is a good first step!

4.4.1.1 Energy Price Projects

The following projects are some that won't reduce energy usage but will reduce the cost of energy.

1. Competitive bids on energy. Other departments such as purchasing could be involved or manage this one.
2. Price hedging. This is a different issue altogether and requires experience and knowledge (and a bit of luck!).
3. Run off-shift or weekends for off-peak costs.

4. Peak demand. Lower peak demand or move peak usage to off-shift peak costs. Sometimes understanding when the peak is occurring can allow for changes to be made without automation. Automation might reduce energy use on HVAC (Heating, Ventilating, and Air Conditioning) or other noncritical equipment when the plant is reaching a peak. Energy storage is beginning to be cost effective.
5. Demand response. Some utilities have programs in which customers are paid if they promise to reduce their demand when requested. High energy-use equipment can be shut down during limited short periods of time.

4.4.1.2 Behavioral Changes

No-cost or low-cost projects maximize the performance of the equipment you have. Figure 4.15 shows how energy intensity varies on aluminum reverberatory furnaces. Low utilization means poor energy intensity.

1. Scheduling and process analysis will maximize utilization. In most furnaces, the more the furnace is utilized, the better the energy intensity. The melting process

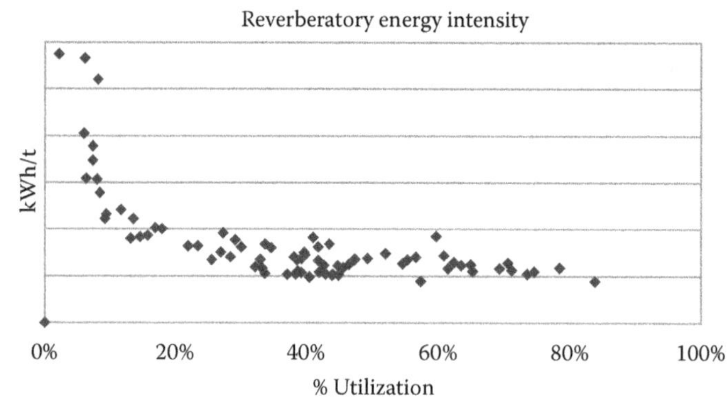

Figure 4.15 Cold air reverb energy intensity.

includes a specific melt time but also includes other steps. The time spent in these other steps should be decreased as much as possible. *Remember, we get paid to melt or heat metal.* Any time we aren't heating metal needs to be minimized. This is also called non-value-added time.

Everything you can do to reduce non-value-added will help. Look for ways to reduce time in

- *Charging.* Charge tables, new staging methods, quicker door open times, remotes in the loader, pay based on pounds charged, larger charging equipment, automated charging conveyors, and so on.
- *Skimming.* Optimize your skimming process. Have equipment ready and ensure there are sufficient pans for all the dross or slag. You don't want to wait for skim pans or find out the skim tool is in poor condition. Test different skimming methods to decrease time while not affecting metal quality. Sometimes we skim too much.
- *Alloying.* Can you change your process to minimize off-analysis product? Can alloys be added with the charge? Is the laboratory close to the process so there is little delay in transporting the sample?
- *Casting.* Are there issues with the next process that cause hold times? Are you casting as quickly as possible? (Steeper trough angle or larger width can help.) Casting quicker may also reduce heat loss so that the bath temperature can be set lower.
- *Scheduling.* Are you running more furnaces than you need? Can a furnace be shut down? Are the right size loads ready? (Small loads can be less efficient.) Can metal be kept hotter by scheduling loads better? Is staging equipment being used effectively? Are

shift changes extending hold times? (Staggered shifts can help.)

- *Transportation.* Are you waiting for a crane, loader, or crucible/ladle?
- *Maintenance.* Is something causing excessive downtime? Can you perform infrared scans to test for developing problems? Are the right priorities being set in maintenance? Are the right spare parts available?

If utilization is low because current sales are low or because there is a designed bottleneck after the furnaces, there are other options to consider:

- Consider shutting down extra furnaces.
- Implement idling practices.
- Investigate lowering the fire rate or power input to the furnace. Most furnaces run more efficiently at a lower rate. This sometimes results in a slower melt rate but a more energy-efficient furnace.
- Move energy-intensive work to off-shifts to save on electricity costs (off-peak costs).

POTENTIAL PITFALL!

Plants don't necessarily respond quickly to changes in production. Running with five furnaces may be great at peak needs, but only three furnaces may be needed at current production.

2. Furnace Utilization, Are you utilizing the furnace correctly? For instance, if you have a stack melter, are you keeping the flue filled with metal? If you have more energy-efficient furnaces and less energy-efficient furnaces, are you maximizing throughput into the most efficient furnaces?

3. Idling, Idle furnaces correctly during weeks, down days, and short hold times.
   a. Use variable frequency drives (VFDs).
   b. On product lines, add limit switches and solenoids to shut off compressed air to blow-offs used to blow off debris. These will shut off the air when the line isn't running.
   c. Document weekend, shift, and break idling of general plant equipment (fans, lights, HVAC, compressed air, etc.)
   d. Optimize furnace idle practices (Belt 2004). Every furnace should have documented idle procedures (Figure 4.16). These idle procedures may be for short periods of holding time such as four hours or for longer periods such as one day or one week. Procedures can involve modifying: fire rate, bath temperature setpoint, roof/flue temperature setpoint, pump setting, covers, blowers, fans, or even proportional integral derivative (PID) control settings. It is best if this can be automated with a push button.

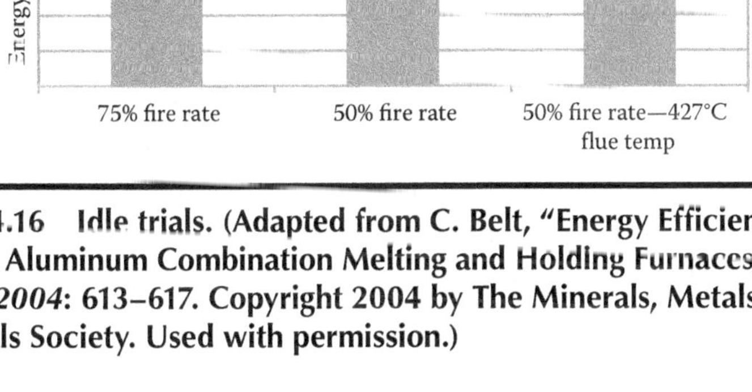

Figure 4.16 Idle trials. (Adapted from C. Belt, "Energy Efficiency Tests in Aluminum Combination Melting and Holding Furnaces," *Light Metals 2004*: 613–617. Copyright 2004 by The Minerals, Metals & Materials Society. Used with permission.)

POTENTIAL PITFALL!

If you have refractory furnaces, do not shut them off and on quickly. Talk to your refractory vendor for the proper shut-down procedures. Each time you take a furnace down and back up will damage the refractory. Every furnace develops large and small cracks in the refractory over time. Cooling the furnace below the melt temperature will mean more damage as the metal expands and contracts within these cracks. Instead, you may want to idle molten aluminum furnaces a little above the melt point at 677°C (1,250°F). If you plan not to use the furnace for three or more weeks, the furnace should be shut down by bringing the temperature down 40°C per hour (75°F). *This time period depends on furnace type. Some furnaces are very rarely cooled down.*

4.4.1.3 Simple Adjustments in Operation

Optimizing the furnace settings can save energy. The following projects are examples.

1. Air/fuel ratio of the burner. There's a reason why this is the first thing listed in many furnace audits. It makes a major difference in energy efficiency, melt rate, and melt loss. If you don't adjust your burners regularly and own a gas analyzer to read the O_2 in the flue gas, start now!
2. Reduce the bath or heat treat temperature setpoint by understanding the needs of the process.
3. Measure and analyze temperatures through the whole process to understand heat loss. For example, covering the troughs will reduce heat loss. The melt furnace bath temperature could then be set lower. Covering transfer pots can also result in reduced bath temperature.
4. Reduce the secondary control setpoint (roof or flue).

5. Reduce temperature overshoot to overcome the flywheel effect due to heat storage in the refractory walls. This could be by optimizing the PID settings, different programming scheme, or better controls.

POTENTIAL PITFALL!

If operators are keeping furnace doors open to bring the metal temperature down, there is something wrong! Analyze the process and controls.

---

6. Reduce high fire. Many furnace burners are specified larger than required as a safety factor. However, metal can absorb heat only at a certain rate. If you exceed that rate, you lose heat up the flue (you may even see flames up the flue). Reducing the fire rate to optimum can save 5%–15% without reducing the melt rate (Belt et al. 2010).

   Test reducing the high fire rate over time. This works best in batch furnaces. As the metal slumps down, the metal surface area decreases. Less heat is absorbed so that the fire rate should be decreased (Belt 2016).
7. Reduce low fire. Low fire settings frequently get out of control. This can especially be seen when you experience temperature creep (when metal temperatures increase even when at low fire).

   Pilot relight is another method of reducing low fire. The pilots on the burners are relit and then the main flame is shut off. This allows a major turndown ratio and is a controls change.

4.4.1.4 Better Maintenance

Make sure you are doing the proper Preventative Maintenance (PM) on equipment. This includes both written forms, performance of the PM, and follow-up.

POTENTIAL PITFALL!

Are all the right checks and inspections being made? Is maintenance taking the inspections seriously? Are problems fixed or just noted and never repaired?

1. Natural gas and compressed air leaks. Compressed air leaks can be found with an ultrasonic gun. A new technology to detect natural gas has been developed in recent years. The camera "sees" in the portion of the infrared spectrum where the gas is visible so that it looks like smoke (Figure 4.17). The camera can scan an area just like taking a movie. This technology is especially useful in the high ceilings of our plants. FLIR is the only company I've found with this camera but several companies can perform the survey. In the surveys I have seen, an average of 1% of the plant natural gas was lost in leaks. Besides the obvious safety aspect, the survey cost was cost justified.

POTENTIAL PITFALL!

A large but hidden gas leak can often be found on the roof. Gas pop-off valves, regulators, and other equipment or appliances are vented to the ceiling or the roof. If a leak develops, it cannot be seen and these leaks are difficult to detect at ground level. These vent pipes should be regularly checked (at least once every quarter) for leaking gas. This can be done with a gas sniffer, an infrared camera, or even by putting a plastic glove on the pipe to see if it inflates.

2. Furnace door open/close time. This is how long it takes to open and close the doors on your furnaces. Over time, performance degrades. If the time can be reduced by 30 seconds and the door is opened and closed every hour, 24 minutes can be saved every day. One internal study showed that for every minute the door was open,

Figure 4.17 Natural gas leak.

the furnace took 2 minutes to reheat to the same temperature. That means you could save 72 minutes in a day!

3. Door seals. Having well-sealed doors improve energy efficiency, melt rate, and reduces melt loss. There should not be holes in the door from flames or external damage (Figure 4.18). Plug the holes or replace the door. Keep refractory repaired in the jambs so heat is not escaping through the edges. Cleaning the ledge should be a part of the operator's job so the door is not held open by dross or scrap metal.
4. Replace motors with high-efficiency motors instead of rewinding motors.

Figure 4.18 Hole in furnace door.

5. Reduce gas used for plant heat. Block off storage areas with flexible door strips or canvas. Install remotes or chains activators for doors to the outside. Use large fans to bring heat to the floor level. Utilize heat tape on piping instead of heating a whole bay to keep water pipes from freezing. Use passive solar or solar heat such as solar roof or wall designs (requires capital).

4.4.2 Capital and Expense Projects

Most projects that save a large amount of energy will take capital or expense money.

4.4.2.1 Proven Technology

Many types of proven energy efficient equipment will save money without extensive testing to prove the concept. The following projects are examples.

1. Lighting upgrades
2. Power factor—adding capacitor banks
3. Air compressor controls

4.4.2.2 Trials

These are projects for which you are unsure of the proper methods, the savings, or the final cost. Try the project in one plant or furnace and then extend the knowledge to other plants after you understand the project better. The following projects are examples.

1. Six-sigma projects. These team projects can do work such as to reduce hold time, improve scheduling, improve charging, and improve scrap mix.

2. Preheat metal charge. This can be done in low-cost "boxes" with flue gases, with zones in heat treat furnaces, or with higher cost stack melt furnaces. In cold weather, even keeping metal stored inside can help.
3. Metal bath stirring increases melt rate and improves energy efficiency.
4. Hydraulic or lubrication upgrades. Better lubrication can reduce energy use with reduced friction. Hydraulic systems can be optimized to reduce energy use.
5. Machining tool and machine upgrades grinding, drilling, lathes, and so on.
6. Burner upgrades. Different types of burners and controls are constantly being developed. The optimum burner depends on the process, so trials are appropriate. Newer styles include pulse, flameless, self-regenerative, and porous.
7. Furnace upgrades. Different burner angles, flue design, ledge design, and other changes can improve energy efficiency without requiring a whole new furnace. Modeling can help determine what changes to make but trials give actual results for your process.
8. Control upgrades. This includes better controls for fire rate and air/fuel ratio.

POTENTIAL PITFALL!

You may have salesmen pushing magnets for your natural gas, fuel, or even combustion air. Even if magnets could "align" fuel molecules, the alignment would be lost before it reaches the combustion point. It's a scam. Don't fall for it.

4.4.2.3 Major Capital Projects

These types of projects cover much more than energy, but make sure the voice of energy is involved. The following projects are examples.

1. New furnaces. More energy-efficient designs are constantly being developed. These include recuperative systems, stack melting, vertical floatation melter (VFM), IDEX scrap decoater and dryer, ultrahigh efficiency cells, continuous heat treat furnaces, and an induction oven with permanent magnets (DOE AMO 2016).
2. New departments. As plants expand with new departments, energy efficiency should be a part of the design process.
3. New plants. It is critical that energy efficiency is a major part of the design of new plants.

4.4.3 Corporate Energy Management Projects

The energy manager may run the projects that individual plants don't have the time to run or projects that cover several plants. The following projects are examples.

1. IT power management—server, computer, and monitor
2. Longer-term research projects—alloy changes, new burner types, and so on
3. Small energy users but at multiple plants
4. Energy training—combustion, compressed air, pumps, motors, and so on
5. Energy policies
   a. Motor management—does your company have a policy on rewind versus new high-efficiency motors?
   b. Requirement to use Best in Class when payback is 75% of the life of the new asset
6. Wind or solar energy. In general, we are in the business of making metal, not producing energy. Solar panels projects tend to be a five- to ten-year payback but can

be good for community relations. However, if you are located in certain areas and have enough land, a wind turbine may have less than a five-year payback. With tax credits, the payback may be even better.

"In open, flat terrain, a utility-scale wind plant will require about 60 acres per megawatt of installed capacity. However, only 5% (3 acres) or less of this area is actually occupied by turbines, access roads, and other equipment—95% remains free for other compatible uses such as farming or ranching. A wind plant located on a ridgeline in hilly terrain will require much less space, as little as two acres per megawatt." (From http://www.awea.org/faq/wwt_environment.html; link no longer available.)

The following two websites give some interesting information on the strength and viability of wind and solar power over the United States:

http://www.nrel.gov/gis/wind.html

http://rredc.nrel.gov/solar/old_data/nsrdb/1961-1990/redbook/atlas/

7. Solar thermal may be a good investment. Similar to pool heater systems, could you use solar heating to warm hot water tanks used for certain processes?
8. CHP systems either produce electricity while the waste heat is used for processes such as plant heat or steam or the waste heat from a process is used to produce electricity. The cleanliness of the combustion gases in the metals industry may cause a problem. CHPs are high cost (some are several millions of dollars) but can provide a good return on investment (ROI) and may make sense in the metals industry. From an online announcement "ArcelorMittal saved more than $100 million by installing a Combined Heat and Power

system to capture 'waste heat' at its steel mill in East Chicago, Indiana."

http://business.edf.org/less-carbon-more-innovation/energy-efficiency-iron-and-steel-industry (link no longer available)

4.4.4 Process Projects

Process projects can produce large energy savings in the materials industry. These projects also tend to provide other cost savings such as better quality or higher throughput.

4.4.4.1 Yield Improvements

Yield improvements can be achieved from not processing material that doesn't go out the door (Belt 2012) (Figure 4.19). The materials industry is known for having high process losses. Typical values for casting are reported to be 64% yield minus an additional 5% product scrap rate (Schifo and Radia 2004).

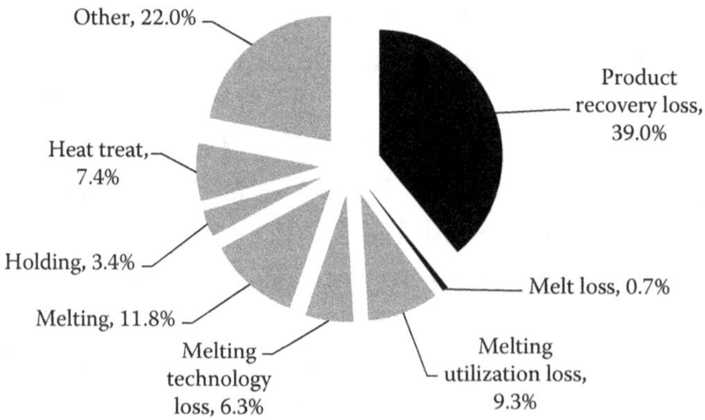

Figure 4.19 Product energy units. (Adapted from C. Belt, "Energy Opportunities in the Aluminum Processing Industry," *Energy Technology 2012: Carbon Dioxide Management and Other Technologies*: 57–64. Copyright 2012 by The Minerals, Metals & Materials Society. Used with permission.)

This is extremely wasteful for energy use because energy is used to process material that is not in the final product.

1. Make sure you are tracking yield! This can be a simple calculation of mass out of the plant divided by mass into the plant.
2. Reduce planned losses. Planned losses include ends (billet, rod, ingot, and coil), alloy changes, casting gates and risers, and so on. Look for creep in design. Sometimes we design in extra metal "just in case." How about controlling the process better instead? Computer modeling is often used to optimize casting design. This includes casting thinner walls but with the same strength and integrity (AMC 2016). Reducing planned losses can also save the energy used to removing the extra material. For instance, not as much electricity is needed to mill a reduced cross section.
3. Reduce unplanned losses. Can over-pours into the mold be reduced? What can you do to reduce off-analysis material? Can a flash-free process be developed? This includes trough "skulls," metal left in crucibles/ladles, and overflow in furnaces or transfer units.
4. Reduce scrap. Cracks, voids, and wrong dimensions can all add to energy lost.
5. Reduce melt loss (metal oxidation). Can you reduce O_2 during the melting process? Can you reduce drops during transfers? Any drop of molten metal in air oxidizes a portion of it. Make sure you charge light metal first with heavy scrap on top or use vortex stirring to pull the metal in quick. Preprocess your chips to remove excessive oil and water and use a vortex to pull the chips quickly into the molten metal. Reducing melt loss does not save much internal energy. However, this metal loss has to be replaced with new metal that may use a large amount of external energy. These energy costs can be seen in the metal cost. Melt loss reduction can be some of the biggest cost savings in our industry!

4.4.4.2 Energy Value Stream Mapping

Like in process mapping, understanding the system helps you see opportunities.

1. Identify process steps to reduce or eliminate. "In some cases, a process heating requirement can be eliminated altogether. For example, there is a current trend to use chemicals that do not require heating to be effective in washing systems used to clean metals parts prior to painting operations" (DOE AMO and IHEA 2015). Many times we reheat metal because of problems in scheduling. Reducing this practice saves energy.
2. Reduce heat loss between processes. This can be using covers for the metal or moving metal more quickly.
3. Use of heat from one process for other areas. For instance, heat generated from air compressors has been used for plant heat.

Energy value stream mapping can be very simple by just recording metal temperatures at different steps of the process (see Table 4.3). Record three types of temperatures at

Table 4.3 Simple Energy Value Stream Mapping

Process	*Expected (°F)*	*Control (°F)*	*Actual (°F)*	*Notes*
Charge material	60	N/A	70	Warm day
Melter	1,350	1,355	1,274–1,349	
Trough 1	N/A	N/A	No TC available	Long trough w/drop
Holder	1,325	1,330	1,329–1,361	Overshooting and door open
Trough 2	N/A	N/A	1,300–1,323	
Trough 3	1,300	N/A	1,299–1,310	

each point: expected (what management or paperwork says the temperature should be), control (what the setpoint is on the controls), and actual (either a direct reading or the reading from the thermocouples).

The results from this batch furnace suggest several avenues to pursue:

- Could the melt furnace control setpoint be lowered? Could the surface of the bath be overheating and more stirring is required?
- Is the holder low fire too high so that metal is overheating?
- Can we reduce heat loss in Trough 2?

Energy value stream mapping can be more complex showing the amount of energy used at each point and the waste heat available (Table 4.4).

The results from this batch furnace suggest several avenues to pursue:

- Why is the melt furnace setpoint so high?
- Why is the casting setpoint above the holding setpoint?
- Melting has the highest waste heat available. Can be used for combustion air preheat, metal preheat, other processes?
- Can hot water from casting be used elsewhere? Or by using a secondary loop with a heat exchanger?

Table 4.5 was used to calculate the approximate percentage of waste heat based on available heat and 60°F combustion air.

4.4.4.3 Other Process Projects

1. Different alloys require different heat treatment. Try to use alloys with the minimum required heat treatment to save energy.

Table 4.4 Complex Energy Value Stream Mapping: Monthly Energy Values

Process	*Setpoint (°F)*	*Gas (MMBtu)*	*Electricity (kWh)*	*Avg. Flue Temperature (°F)*	*Waste Heat (MMBtu)*	*Other Losses*
Dryer	250	1,021	2,356	400	159	
Melters	1,400	37,924	212,058	1,800	20,081	4% Dross
Holders	1,330	5,398	108,390	1,500	2,426	
Casting	1,350	0	12,441	150	2,921	0.5% Scrap

Table 4.5 Flue Loss

Flue Temperature (°F)	*Flue Temperature (°C)*	*Flue Loss (%)*
100	38	7.6
200	93	10.2
300	149	12.9
400	204	15.6
500	260	18.3
600	316	20.9
700	371	23.6
800	427	26.3
900	482	28.9
1,000	538	31.6
1,100	593	34.3
1,200	649	36.9
1,300	704	39.6
1,400	760	42.3
1,500	816	44.9
1,600	871	47.6
1,700	927	50.3
1,800	982	53.0
1,900	1,038	55.6
2,000	1,093	58.3
2,100	1,149	61.0
2,200	1,204	63.6
2,300	1,260	66.3
2,400	1,316	69.0
2,500	1,371	71.6

2. Off-specification metal can add to melt times and energy use. Work toward alloying correctly the first time and every time.
3. Use of scrap materials versus prime material saves embedded energy. For instance, primary aluminum used 26,000 Btu/lb (16,760 kWh/t) while the secondary aluminum process uses only 2,200 Btu/lb (1,420 kWh/t) (Baldwin 2015). This energy savings is external to the plant but results in lower material costs.
4. Eliminate major steps in the process. For instance, casting metal directly into sheet, thin slabs, or rod saves energy intensive steps. Direct reduction ironmaking eliminates the iron ore agglomeration and cokemaking processes. Electric arc furnace steelmaking does not use the iron ore agglomeration, cokemaking, or ironmaking processes. Ladle furnaces in steelmaking can be replaced with a continuous series of vessels (DOE AMO 2015).
5. Purchasing molten metal directly from the vendor versus solid metal (sows, ingots, bars) saves the energy of remelting the metal. Over-the-road methods of delivering metal can safely transport molten metal several hundred miles.
6. Investigate utilizing near net shape methods in casting (better surface, less distortion, etc.).
7. Additive manufacturing can make many final products using much less energy (Baldwin 2015).
8. Semisolid processing such as thixoforming or rheocasting means that extra energy is not used to melt the metal fully.

4.4.5 Public Relations

Some projects do not make strict financial sense, but could be done to highlight the program with employees or the community.

1. Solar panels.
2. Leadership in Energy and Environmental Design (LEED) certification appears worthwhile, but buildings are a small part of our business.
3. Compact fluorescent light bulb (CFL) handouts to employees.

POTENTIAL PITFALL!

Sometimes you don't have money available for energy projects. Many small projects can be better than waiting for capital on big projects. For instance, having each plant measure the time it takes to open and close the furnace doors, compare, and improve can provide major savings for little money.

4.5 Process Heating

Energy used in process heating is by far the main use of energy for our industry. It is important to understand how our furnaces work to transfer heat to metal and where we lose heat.

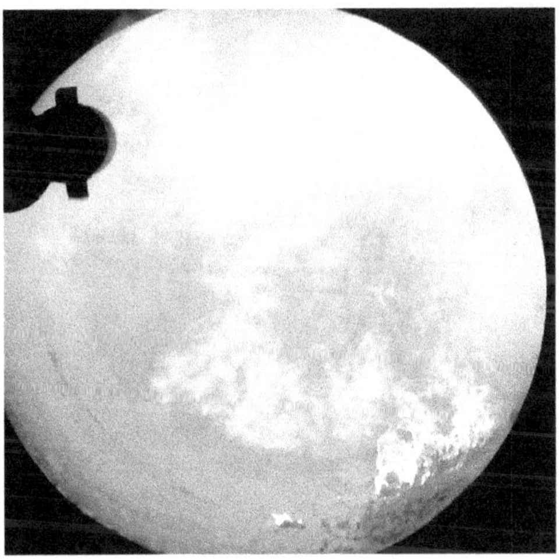

4.5.1 Sankey Diagram

Sankey diagrams are a specific type of flow diagram, in which the width of the arrows is shown in proportion to the flow quantity in a heat balance. They are typically used to visualize energy or material transfers between processes. In this case, we are looking at the heat going into and out of a furnace.

It is important to understand each input and output to optimize the system. The PHAST example in Figure 4.20 came from the Department of Energy (DOE)'s PHAST program (Process Heating Assessment and Survey Tool). The DOE has updated to an online tool called Process Heating Modeler Tool (PHMT).

4.5.2 Gross Fuel Heat Input

Producing energy or heat from fuel can be optimized.

1. Better fuel mixing or reduction of incomplete combustion may be possible in some situations. For instance, combustion can be incomplete if cold metal is in front of the burner.
2. Are your burners in good shape? If you have carbon buildup or metal splash covering your burner, you may not be getting the air and gas you think you are getting. Burners are designed to mix the air and gas at certain locations in the flame. Soot buildup makes the burner less efficient.
3. Dehumidifiers have been used in some furnaces such as in a cupola so that heat is not used to heat the moisture in the combustion air. Benefits have included energy efficiency, higher melt rate, reduced flux, and more consistent operation.
4. Improved coke quality in blast furnace ironmaking reduces the amount of energy required (DOE AMO 2015).

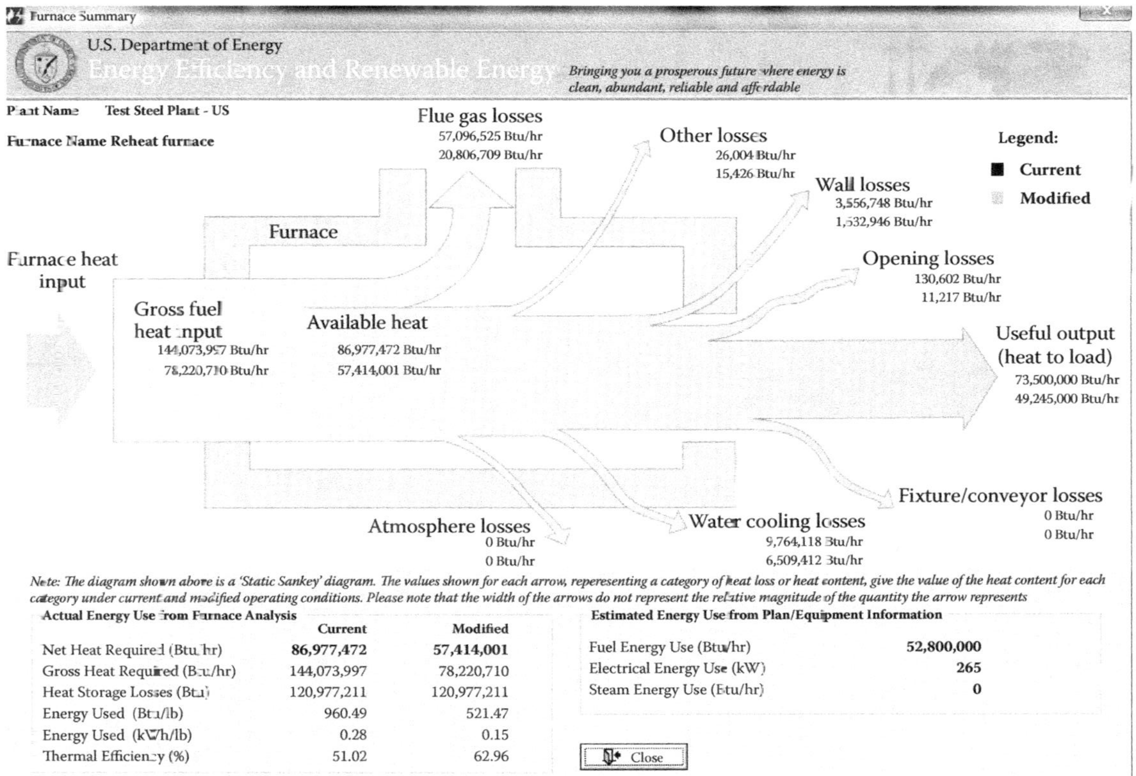

Figure 4.20 PHAST example.

4.5.3 Useful Output (Heat to Load)

The transfer of heat to the load can be improved.

1. Stirring brings the hot metal at the top surface down and brings cold metal to the surface for better heat transfer.
2. Residence time
   a. Understand the flow of hot gases in your furnace. The longer the gases are in the furnace, the more the heat is transferred to the charge or to the refractory.
   b. If the flue is too close to the burner or directly across from the burner, you may be short-circuiting so the heat doesn't have time to transfer to the metal. Short-circuiting occurs when gas directly moves from the burner to the flue without transferring heat to the metal. If metal is piled up in front of a burner, it may also short-circuit the flames.
   c. Furnace design affects residence time. In general a double pass (or W design) is better than a single pass (Figure 4.21). A small thing like lowering the flue if the flue is on the roof of a furnace or at the top of the wall will slow down the loss of heat. However, modeling can help here. With some furnace designs, raising the flue opening is better to reduce short-circuiting. It depends on your particular furnace design.
   d. Other methods to increase residence time include lowering the fire rate, modeling the flame to optimize the

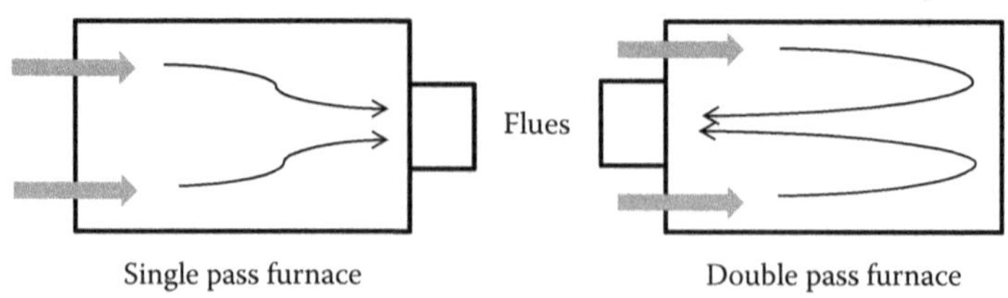

Figure 4.21 Furnace design.

burner angles, adjusting the burners to different settings to swirl the gases, and adding a charge section to preheat metal.

4.5.4 *Flue Loss*

Most of the energy is lost up the flue.

1. Air/fuel ratio. Optimizing the air/fuel ratio makes a major difference in energy efficiency, melt rate, and oxidation of metal. This ratio is so important that a special section is needed to discuss the topic.

Air/Fuel Ratio

The chemical formula for methane (the primary component in natural gas) and oxygen combustion is shown in Equation 4.1. This means a 2:1 stoichiometric ratio of oxygen to methane fuel.

$$\begin{aligned} &2O_2 + CH_4 \rightarrow CO_2 + 2H_2O + \text{heat} \\ &2{:}1 \end{aligned} \tag{4.1}$$

Air is composed of very roughly 80% nitrogen and 20% oxygen. This gives us Equation 4.2. The ideal ratio for air and methane is approximately 10:1.

$$\begin{aligned} &8N + 2O_2 + CH_4 \rightarrow CO_2 + 2H_2O + 8N + \text{heat} \\ &10{:}1 \end{aligned} \tag{4.2}$$

The stoichiometric ratio for natural gas depends on the ratio of the gas chemistry (methane, ethane, propane, etc.). These other fuels require a different ratio than the 10:1 for methane. However, pipeline natural gas is 95%–98% methane in the United States (EPA 2016) so the stoichiometric ratio for natural gas is normally listed as 10:1.

Stoichiometric ratio is the perfect ratio. The actual ratio the burner is running makes a major difference in performance.

- A low air/fuel ratio (rich gas) means there is not enough oxygen to burn all the fuel, so less energy is generated. In this case, CO is found is the flue gases.
- A high air/fuel ratio (lean gas) means too much air comes along. This cools the flame and lowers the available heat. In this case, extra O_2 is found is the flue gases.

In many furnaces, the optimum setting means 1%–2% O_2 in the flue gases. This equates to 5%–10% excess air. The excess air is needed to burn all the fuel in a less than perfect furnace. This optimum setting varies by process. Some processes tend toward a rich gas mixture so that O_2 is low and reduces metal oxidation. Some low-temperature processes tend toward lean, so more air pushes the heat to dry product.

Watch that you are using the correct terminology for air/fuel settings. Excess air and flue gas oxygen mean something different!

Stoichiometric ratio (SR) = The ratio of oxidant to fuel required for perfect combustion

Example: 10:1 for air to natural gas or 2:1 for oxygen to natural gas

Air/Fuel ratio (a/f) = Air volume/Fuel volume

Example: 10.5:1 air to natural gas.

Excess air (XSAir) = Percentage of additional air added over stoichiometric combustion

Example: 10.5:1 air/fuel ratio equates to 5% excess air.

$$(\text{a/f} - \text{SR}) \div \text{SR} = \text{XSAir} \times 100\% = \%\ \text{XSAir} \tag{4.3}$$

Air/fuel example:

$$(10.5\text{ a/f} - 10.0\text{ SR}) \div 10.0\text{ SR} = 0.05\text{ XSAir} \times 100\% = 5\%\text{ XSAir} \quad (4.4)$$

Oxy/fuel example:

$$(2.02\text{ a/f} - 2.0\text{ SR}) \div 2.0\text{ SR} = 0.01\text{XSAir} \times 100\% = 1\%\text{ XSAir} \quad (4.5)$$

Flue gas oxygen = Percentage of oxygen within the flue gas composition by dry volume

Example: 10% excess air translates into approximately 2% O_2 in the flue.

Many flue gas analyzers will read flue gas oxygen and calculate the excess air value. These analyzers should also read CO in the flue. CO means there was not enough available oxygen to complete the normal reaction, poor mixing, CO_2 dissociation in the high temperature zones, or that something like cold metal stopped the reaction.

Flue gas analyzers are not very expensive (may cost less than $1,000). They need to be calibrated regularly because the cell life is relatively short. The analyzer is connected by a rubber hose to a ceramic or stainless steel tube. The tube is inserted in the flue gas and left there for 20–30 seconds. The following are some issues:

- The furnace should be stable and at high fire with positive furnace pressure. Negative pressure means that air pulled in at openings will affect the reading.
- The sample should be taken where combustion is complete. The flue is better than the roof area since combustion is still occurring inside the furnace box.
- Care should be taken to ensure ambient air does not affect the reading such as air dampers. A good location is 0.5 meter in from the top of the flue with any air dampers turned off.

Some burner control systems are considered cross-connected. In this system, the air is controlled by a motor and linkage. A small combustion air pipeline brings air pressure to a regulator in the gas line so that more air means more gas. The gas is cross-connected to the combustion air. This method has inherent variations. A better method of controlling the air/fuel ratio is called mass flow. Mass-flow controls measure the flow for both air and fuel. The air/fuel ratio is controlled by computerized controls. Multiple issues need to be addressed to optimize the air/fuel ratio over time in both types.

- Variation occurs in both pressure and temperature in the fuel and air that affects both types. This variation can occur between seasons and even during a single day. Mass-flow control with pressure and temperature compensation is more accurate.
- Linkages and regulators has bias, nonlinear response, and hysteresis effects that result in less than perfect air/fuel ratio as the burner moves from low to high fire and over time (Lampe 2016). Mass flow can adjust flows better.
- Air and gas filters/strainers can plug, which reduces the air flow and changes the ratio. Metal splash and carbon build up (soot) in the burners can change the ratio.

All this variation means a less-than-perfect air/fuel ratio. Periodically using a flue gas analyzer to measure O_2 and CO is good, but mass-flow burner control is better overall.

There can be a variation over time in the heat content of the fuel, requiring a different air/fuel ratio. Oils and coatings can provide fuel during the first part of the melt cycle that again affects the ratio. Other controls can overcome these issues. Sensors can monitor the O_2 and/or CO in the flue and adjust the air/fuel ratio to optimize the system. Many types are available on the market (Boltz and Boltz 2015). These sensors can be a maintenance issue, but many companies have had success with flue gas sensors.

2. Reduce flue or roof temperature setpoint. This simple setting change may improve energy efficiency without affecting melt rate. A side benefit is reduced maintenance to the furnace flue and stack. Testing is the best way to know the optimum temperature setting.
3. Optimize the high fire rate (see Figure 4.22). Many furnaces are overfired. In many cases, reducing the fire rate can improve energy efficiency without reducing the melt rate (Belt et al. 2006). This will reduce heat loss in the flue and increase residence time. Higher residence time means there is more heat transfer to the load.

   In some processes, it may be best to reduce the fire rate after a set time or condition. For instance, batch melting may take two hours. Reducing the fire rate after one hour can improve efficiency because the metal surface area has now decreased.

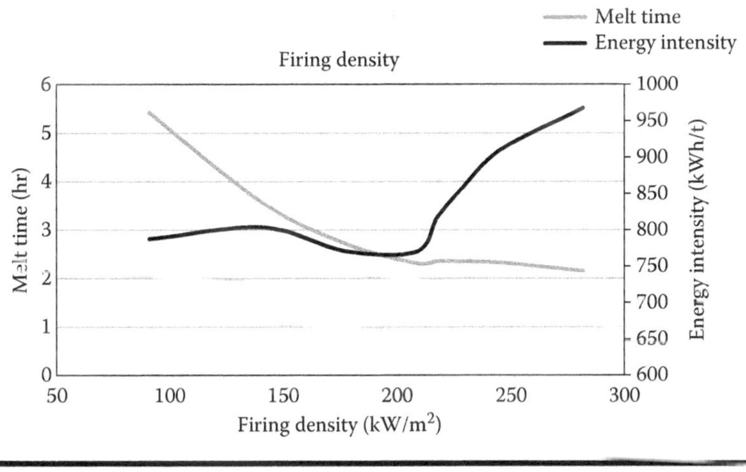

Figure 4.22 Melt time and energy intensity versus firing density. (Adapted from C. Belt, B. Golchert, P. King et al., "Industrial Application of DOE Energy Savings Technologies to Aluminum Melting," ***Light Metals 2006*****: 881–885. Copyright 2006 by The Minerals, Metals & Materials Society. Used with permission.)**

4. Oxygen burners or oxy-fuel burners reduce the amount of energy used because nitrogen in the air isn't heated.
   a. Total oxygen enrichment can save up to 35% on energy usage.
   b. Lower than 100% O_2 can be more cost efficient.
   c. Pros: Higher melt rate along with cooler and lower flow rate of flue gas to the baghouse.
   d. Cons: Can be higher melt loss. Many companies have seen damaged refractory due to the high flame temperature when oxygen enrichment is first installed.

POTENTIAL PITFALL!

Make sure a full analysis is done. The cost of oxygen can offset the energy savings. It depends on the local cost of oxygen.

---

5. Reuse of energy to preheat combustion air.
   a. Low level preheat. Use inside air, hot air from the ceiling, or air from under the furnace to bring to 100°F–250°F (38°C–120°C). Can save up to 5%.
   b. Recuperators. Preheat air using tubes inside the flue, a heat exchanger surrounding the flue, or external heat exchangers such as heat pipes. Air can be brought up to 900°F (480°C), with energy savings of up to 25%. However, flue gases in many metal furnaces can quickly destroy heat exchanger materials owing to the high temperature and chemical composition.
   c. Regenerative burners use flue gas to heat a special media bed. Combustion air can be preheated up to 1,800°F (980°C), saving up to 45%. Note: These systems can increase maintenance costs.
   d. Preheated combustion air may also increase melt rate due to the hotter flames.
   e. Watch for NO_x levels because NO_x formation increases as the flame temperature increases. Low-NO_x burners may be required.

6. Watch for excessive fume collection. Excess heat can be drawn out of the flue or from the door area if the exhaust or baghouse blowers are set too high.

4.5.5 Wall Loss

Heat is lost through the walls, floor, doors, and roof.

1. Optimize refractory design to reduce heat loss on furnace exterior. Use programs such as Hauck's E-solutions or talk to refractory companies.
2. Decrease the emissivity of the walls to reduce radiation heat transfer. A silver paint is best. Consider cleaning the walls every few years when they get dark and gritty.
3. Reduce airflow along walls to reduce convection heat transfer. There should be no man-fans blowing on the walls. A broken up surface such as with structural beams will reduce the airflow and heat loss.
4. Don't forget the heat loss through the doors. Make sure these are properly insulated and maintained.
5. Do most of your improvements on the upper walls and roof. This is where the inside temperatures are the hottest and more heat is lost. The pie chart in Figure 4.23 is from an aluminum front charge furnace.

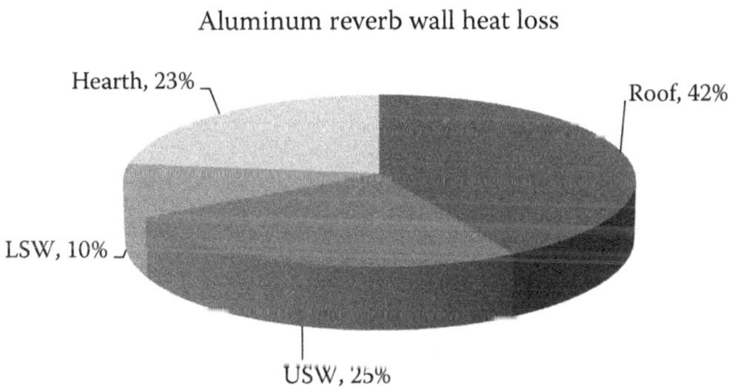

Figure 4.23 Heat loss from aluminum melt furnace. LSW = lower side wall; USW = upper side wall.

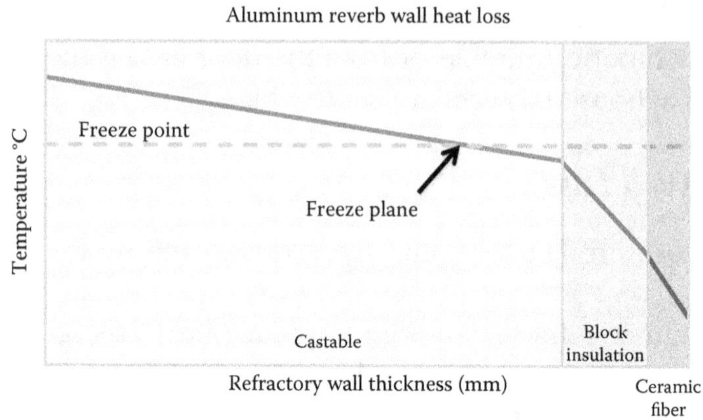

Figure 4.24 Aluminum reverb refractory temperature profile.

POTENTIAL PITFALL!

Too much insulation can change the position of the freeze plane. The freeze plane is the point where the temperature is same as the molten metal point (see Figure 4.24). Moving the freeze plane into the insulation layer could result in damaging the furnace structure or allowing a metal spill. Talk to your refractory vendor!

4.5.6 Opening Loss

Any fixed or variable opening (Figure 4.25) will result in heat loss.

1. Reduce door open time (charging, skimming, alloying, furnace cleaning, temperature readings, dome or door closing time, etc.).
2. Close any openings (e.g., stuff fiber blanket around thermocouple openings). For crucibles or ladles, lids could be made or some plants turn the ladle upside during the heating process.

Figure 4.25 Furnace door opening.

3. Lower openings. The higher the opening is in the furnace, the hotter the air and the higher the pressure, which means more heat loss.
4. Improve the furnace pressure system.
   a. When the furnace pressure is too high, heat can be lost as flames out the door. This also reduces door and refractory life.
   b. When furnace pressure is too low, cold air can infiltrate the furnace. This cools the flame and the load. It also causes incomplete combustion. Extra air can also increase melt loss.
   c. Preventative maintenance on the damper (Figure 4.26) and furnace pressure transmitter is important!

POTENTIAL PITFALL!

Many times operations believe the furnace pressure system is working fine. Try opening the door and reading the furnace pressure. It should read zero because it is reading atmospheric pressure. Then close the door again. Run the furnace at high fire and low fire to see if the furnace pressure reaches setpoint. Another way is to take a strip of paper and *carefully* place it near a small opening in the door. If it pulls toward the furnace, the furnace pressure is negative. If it pushes out, the furnace pressure is positive.

Figure 4.26 Furnace damper.

4.5.7 Conveyor Loss

In some furnaces (particularly heat treat) racks, trays, or conveyors transport the product. Energy is used to heat this conveyor.

1. Reduce the weight of the rack or tray.
2. Keep the rack or conveyor hot between uses by either charging quicker or covering the rack.
3. Dirty scrap can be seen as conveyor scrap. If you have sand carryover in iron castings, it requires twice as much energy to heat a pound of sand versus a pound of iron.

4.5.8 Other Losses

The following are some miscellaneous heat losses.

1. Water cooling loss. Heat lost through water cooling systems in the doors or other locations
2. Atmosphere loss. Heat required to heat an added atmosphere such as nitrogen

3. Hot surfaces other than refractory or walls
4. Heat content of coal ash, slag, dross, or other byproducts

4.6 Waste Heat Recovery

The metal industry is full of sources of waste heat (Thekdi and Belt 2011) that could be used instead of being lost to the environment. These sources include

1. *Gases*. High-temperature flue gases (melting, calcining, sintering); low-temperature flue gases (heat treat, dryers, regens, emission control); low-temperature air from direct and indirect cooling (product or equipment)
2. *Liquids*. Cooling water, wash water, blow down water, steam leaks, condensate
3. *Solids*. Hot product, heat of reaction for processes (dross, slag), hot surfaces (furnaces or conveyors)

POTENTIAL PITFALL!

Not only do we waste heat in these processes, but we may also use more energy for cooling. For instance, we use fans for cooling towers or to directly cool product.

4.6.1 Uses

Instead of wasting this heat, there are different methods to utilize the heat. In order of cost-effective solutions:

1. *Reduction.* Optimizing your system should always be the first step. It is more efficient not to have as much waste heat versus reusing the waste heat.
2. *Recycling.* Use waste heat energy within the source itself.
    a. Combustion air preheating—low-temperature, self-recuperative, recuperative, regenerative burner systems
    b. Load or charge preheating can be used to preheat or remove moisture from charge material. The difficulty depends on type of material and available space
    c. Make-up air or water preheating
    d. Considerations—compatible with variation in operating systems, 5%–40% improvement in energy intensity, may increase melt rate and improve melt loss, and typically for 200°C (400°F) or higher temperatures
    e. Typical payback of 1–4 years
3. *Recovery.* Use waste heat energy for other systems within the plant or area.
    a. Heat water
    b. Steam generation
    c. Plant air (HVAC) heating. Sources can be from flue gases, hot water, or hot metal
    d. Absorption cooling system
    e. Dry coal for cokemaking process
    f. Cascade use—use for lower temperature process

    Considerations: It is difficult to match heat supply to heat demand, typically for 120°C (250°F) or higher, may require heat exchangers to transfer heat to clean air or another medium, and moderately expensive because heat needs to be transferred longer distances.

    The typical payback is one to six years.

4. *Recovery—Electrical.* Use waste heat energy to produce electricity.
   a. Conventional steam system—boiler, steam turbine, generator
   b. Organic Rankine Cycle (ORC)
   c. Ammonia–water system (Kalina, Neogen)
   d. CO_2 power cycle
   e. Combined heat and power (CHP) is a special version of this. These units either use fuel to make electricity and then the waste heat is used within the plant or use waste heat to make electricity. CHP use is high in Europe and is growing for places such as hospitals and hotels, where CHP is used instead of steam boilers. The use is growing in the metal industry
   f. Thermo-electric generation (TEG) systems are in an early developmental stage and their use cannot be economically justified at this time except in special cases. The reported efficiency has been less than 5% and they are expensive.

   Considerations: A continuous or predictable flow is needed, a source above 315°C (600°F) is preferred, and the waste heat source must be relatively clean and free from contamination.

   The typical payback is longer than five years.

4.7 Furnace Testing

There are at least three different types of furnace tests (Belt 2014). Each has its time and place.

4.7.1 Furnace Benchmark

These are tests run to commission a furnace. In most cases, specific metrics are being tested. For a new furnace that is being tested before the final buy-off, those metrics need to be defined in the purchase order. For instance, if the metric is kWh per ton, is it during one specific batch or during a week of normal production? What type of material or batch size will be tested? Whenever a furnace is started new or after a major rebuild, running a test to understand energy intensity, melt loss, or melt rate is very important. It is good practice to have a written procedure for these tests so there are no questions afterward.

POTENTIAL PITFALL!

Melting energy intensity is normally very different from energy intensity over a day, week, or month. Normally, hold times will affect the value.

4.7.2 Furnace Process Heating

Tests are run to understand the heat balance of a furnace such as a Sankey diagram. These tests will measure energy intensity, but also try to understand the heat losses by measuring outside wall temperatures, flue gas temperatures, metal temperatures, and so on. These tests are frequently run during a process heat assessment or can be run using the DOE's PHAST program.

4.7.3 Furnace Operation

While the heat losses are very important, understanding the furnace process can provide insight to improve the basic production methods. Measurements should be taken of gas readings, charge weight, cast weight, dross/slag weight, and times for every event (charge, burner on, door open, door close, casting, maintenance downtime, etc.). A test like this should monitor the process for 8–12 hours.

Running a single test on a furnace is good, but running tests on different shifts, different furnaces, or different plants is much better. Comparing performance gives you a good idea of changes that will improve the process by learning from the best. It is preferred to have the same person monitoring all the tests so that the results are comparable. That person will also be able to note less tangible differences between furnace operations.

It is critical to have a standard procedure for these types of tests. This would include standard charge materials, charge amounts, and taking the same measurements at the same time in the procedure. Keeping plenty of notes during the test will also help.

The final report on these tests could include energy intensity, melt rate, melt loss, average batch size, door open time, skim time, chemistry check time, casting time, and maintenance downtime and reason. Recommendations should be made for improvements.

While your normal data collection may tell you that there are differences in pounds melted between shifts, this type of testing may show how better communication improves charge time. Normal data collection may show a problem with energy intensity, but this type of testing may prove that problems with flame-outs are causing excessive downtime. A comparison between plants may show that throughput is higher at one. Running these tests may show something simple like charge buckets were bigger at one plant or that a cramped plant layout is slowing charge times at the other.

4.8 Technical System Analysis

Not all the energy used in plants is used in process heating. Other top users are compressed air, fans, pumps, motors, and HVAC. You don't need to be an expert in these fields, but some knowledge helps. It also may help if one person at your company becomes the expert in these fields. He or she can help different plants to optimize their systems and become the "go-to" person for the topic.

The following are some comments on the various areas, but is by no means a definitive list.

1. *Compressed air.* Compressed air is used in almost all metal industry plants. A single compressor of 250 HP (190 kW) can cost $150,000 per year in energy and more in maintenance. Savings can be found in two areas.
   a. *Demand.* This is how air is used. Before you worry about buying a new compressor, this should be your first step. The goal is to use the least amount of air and at the lowest pressure.
      i. The biggest and least costly savings is repairing air leaks. If your plant doesn't have a leak program, start one. A good plant will still have 10%–20% air leaks because fixing every leak is just not cost effective. It is not unusual (but still bad!) to see 50% of the air lost in leaks. The easy method to understand your leaks is to see how much air is used during a down day. A single air leak that you can hear wastes over $700 in a year.
      ii. Add low cost solenoids and timers on blow-offs so that air is used only when you need it.
      iii. Use modern air nozzles that entrain air. These can use as low as 10% of the air of an open pipe or those "nozzles" made by hammering a copper pipe.
      iv. Look at all your air use to see if it makes sense. Compressed air is expensive. Other options can normally be cost-justified quickly.
         A. High-pressure regenerative blower for blow-offs
         B. Motorized pump instead of an air pump
         C. A fan instead of an open air line to cool people

D. A vacuum cleaner or brush instead of an air hose to blow off machines or people. *Using compressed air to blow-off metals chips on a person is dangerous!*

v. Reduce the plant air pressure to the lowest possible. For every 2 psi you can lower the compressor you save 1% on energy.

vi. Do you use zero air loss drains where condensate forms (receivers, etc.)?

b. *Supply.* This is how the compressed air is made. The goal is to produce air at the least cost.

i. Investigate the controls on your compressor. Just having the right settings can help so compressors aren't "fighting" each other. It is worth bringing in an outside expert to review this.

ii. Use cool outside air as the intake to the air compressors.

iii. The use of air storage units (receivers) helps even out air production and can prevent the start-up of another compressor.

iv. How old are your air compressors? Does a new energy-efficient compressor make sense?

Take a class or read up online at sites such as www.compressedairchallenge.org or www.energy.gov/eere/amo/compressed-air-systems.

2. *Pumps.* Pumps are used extensively for coolant systems and cooling towers. Energy is wasted when pumps are oversized or bypasses are always in use. More information is available at www.energy.gov/eere/amo/pump-systems.

3. *Fans and blowers.* Fans are used to remove heat. Blowers are used in locations such as combustion air. One quick method to save is to make sure fans are off when they aren't needed (e.g., roof fans during the winter). Correct sizing of the blower and motor is important. Learn more at www.energy.gov/eere/amo/fan-systems.

4. *Motor management and drives.* Motors are used everywhere in our plants.
   a. Consider high-efficiency motors when buying new.
   b. In most cases, it is better to buy a new motor versus doing a rewind. Let your motor supplier run the numbers.
   c. Variable frequency drives (VFDs) will save energy on motors with variable load.
   d. Use cogged belt drives when appropriate.

   More information is available at www.energy.gov/eere/amo/motor-systems.
5. *Steam.* Steam boilers are used for heat in many processes. Read more at www.energy.gov/eere/amo/steam-systems.
6. *Lighting systems.* Lights may only be 1%–2% of your energy use, but new lighting makes sense. An extra benefit is that many electric utilities have rebates available for new lighting. Most lights have a simple ROI of one to four years, with even better returns with financial help from the electrical company.
   a. The typical conversion is from high-pressure sodium and metal halide lights to fluorescent T5/T8 lights. In many cases light-emitting diode (LED) lights are cost justified depending on current equipment costs and rebates.
   b. Energy savings of 50% are possible.
   c. Quick relight capabilities will improve safety.
   d. A longer life of 50% and low replacement costs reduce maintenance costs.
   e. New lighting should retain better luminosity over time and provide good color. These qualities help with product inspection.

POTENTIAL PITFALLS!

Ceiling height and temperature (both cold and hot) are critical in selection of lights. Get help from vendors in selecting the best lighting in the application.

Investigate both bulb life and lumen maintenance for the lighting you investigate. Metal halide is known to lose much of the luminosity over time. T5 and T8 are much better. LED lighting varies in performance.

7. *Fuel management.* Energy is used both internal and external to plants for transportation. Savings are possible.
   a. Reduce transfers. Loads could be larger, distances shorter, or transfers can be eliminated.
   b. Reduce truck idle times.

c. Automate transfers. Small computerized in-plant units use less energy.
d. Investigate alternative fuels (biofuel, plastics, oils, coke oven gases, etc.).

Technical assessments. Plant or system assessment or audits are undertaken to identify energy performance improvements with a formal written report as an output. Individual components or subsystems might not be addressed with equal weight, but system assessments should be sufficiently comprehensive to identify the major energy efficiency opportunities for improving overall system energy performance.

Assessments enable you to

a. Identify opportunities for waste reduction and energy improvement
b. See best practices in action and to document them
c. Involve facility staff in learning and knowledge transfer
d. Prioritize projects and track progress
e. Save on energy costs and improve your bottom line

The following resources can help in running an audit.

BS EN 16247-1:2012 Energy Audits
This European standard defines the attributes of a good quality energy audit. It includes methods and deliverables for general energy audits.
ASME EA-1-2009 Energy Assessment for Process Heating Systems (www.asme.org)
This standard covers audits of process heating systems. It describes the aspects of a quality process heat energy assessment. There is both a standard and a guidance document.

See Appendix I for resources available for both internal and external assessments.

Chapter 5

Do the Improvements

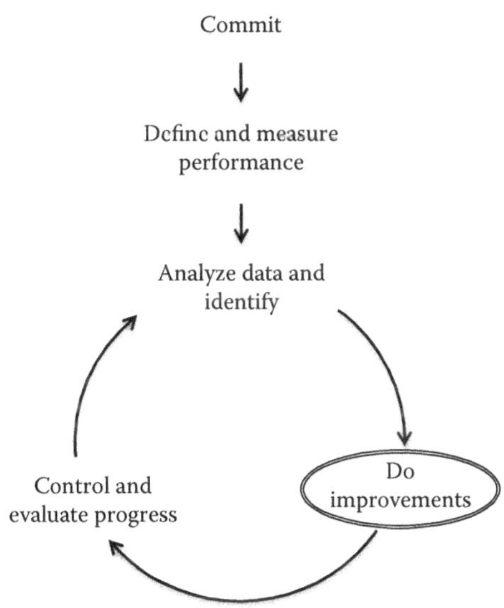

5.1 Creating the Action Plan

You've come up with all types of ideas on projects you can do. Now it's time to make a list, evaluate these ideas, and then find a method to accomplish the projects.

5.1.1 Potential Project List

Document opportunities on a project list. Lists may be at the department, plant, or company level. Table 5.1 shows a short example.

1. Define target energy performance.
2. Calculate potential energy savings (see Equation 5.1).

$$\begin{aligned}\text{Annual savings} &= (\text{Current energy intensity} - \text{Goal energy intensity}) \\ &\quad \times \text{Annual mass} \times \text{Energy cost} \\ &= (805 - 750)\,\text{kWh/t} \times 50{,}000\,\text{t/yr} \times \$0.10/\text{kWh} \\ &= \$275{,}000/\text{yr}\end{aligned} \tag{5.1}$$

3. List and quantify other benefits. Most energy projects have other benefits—melt loss, melt rate, quality, labor, maintenance, environmental.
4. Find potential solutions to reach goal. Very roughly estimate the cost for potential solutions. This may be labor, expense, or capital.
5. What is the business case? Calculate simple return on investment (ROI).

$$\begin{aligned}\text{ROI} &= \text{Cost} \div \text{Savings} \\ &= \$100{,}000\ \text{Cost} \div \$275{,}000\ \text{Annual savings} \\ &= 0.36\ \text{ROI}\end{aligned} \tag{5.2}$$

6. Roughly define resources needs (internal or external manpower).
7. Refine the list to determine priorities.
   a. It may be ROI or it may be high savings. When capital is not available, low-cost projects may be best. Or it may be best to tie in to other projects that are already planned so they can be done at the same time. Some projects may need to be delayed until manpower is available.

Table 5.1 Potential Project List

Department	*Equipment*	*Description*	*Est. Savings*	*Est. Cost*	*ROI (yr)*	*Comments*
Casting	Melt furnace	Bring to benchmark	$275,000	$100,000	0.36	Six-sigma project
Casting	Crucibles	Self-recuperative burner	$3,250	$7,500	2.3	Add TC during lid reline
Plant	Capacitor bank	Improve power factor	$28,700	$32,000	1.1	Upgrade cap. bank
Plant	Lighting	Upgrade to T5	$70,000	$150,000	2.1	Investigate utility rebate

b. Don't underestimate the politics involved. Sometimes a visible smaller savings is better than a larger hidden savings. For instance, upgraded lighting is something everyone can see.
c. Look for alternate financing, grants, incentives. These can be found from your energy supplier, Energy Service Company (ESCO), or http://www.dsireusa.org/. If you haven't talked to your utility companies, do this now! More rebates are expected by utilities as government policies change.

Remember, it's all about money when making your case. Periodically let your management know about the potential cost savings from these energy projects. Writing monthly, quarterly, or annual reports on plant energy status and potential projects is a good method. Keep bringing up potential energy projects to management. Remember, a plant manager may be worried about a safety issue, a labor issue, hiring a new supervisor, and planning the next meeting with his or her boss when he or she reads your report the first time. Energy is not the manager's first priority. Especially watch the internal cycle of projects. For instance, if the plant works on next year's budget in August, send a list of potential projects in July. Management will be looking for ideas at that time. If you ask for capital in February, the annual budget is probably already set.

Before a project is approved, more work will be needed to refine these numbers along with using the appropriate accounting method for your company (ROI, rate of return, life-cycle cost model, etc.). However, using rough numbers helps prioritize potential projects.

POTENTIAL PITFALL!

Periodically check your assumptions. As you identify new methods or obtain better quotes, keep the project list up to date. Priorities will change over time.

5.1.2 Active Project List

After you've started work on projects, you need to keep track of them. This is a higher level list to keep track of projects. It's good to keep completed projects in this file to transfer the knowledge. Table 5.2 shows an example.

1. It could be within company's Continuous Improvement program or a separate program.
2. Assign the person responsible.
   a. Internal roles. Who should be involved and what are their responsibilities?
   b. External roles. Will you use consultants, vendors? If so, to what degree?
3. Assign a completion date.
4. Develop effective performance goals.
5. Include comments or scope.
6. Determine roles and resources.

5.1.3 Project Status File

Many energy projects will take months or even years to complete. It's important to monitor the status and later confirm savings. Table 5.3 shows an example.

1. Establish a system to track progress of projects.
2. Update regularly (e.g., monthly or quarterly) to reflect recent achievements, changes in performance, and shifting priorities.
3. If possible, have the project leader update the status. Updating the status and/or team meetings tend to cause more work to be accomplished, especially right before the update is needed. No one wants to be embarrassed by no work being done.

Table 5.2 Active Project List

Project	*Leader*	*Due Date*	*Goal*	*Projected Savings*	*Comments*
Compressed air leaks	Jerry Martin	7/1/14	Reduce leaks 1200 → 600 cfm	\$37,500	Record weekend use. Leak program.
Six-sigma melt furnace	Jane Orville	3/1/14	2100 → 1800 Btu/lb	\$285,000	Team members: ....

Table 5.3 Project Status File

Project	*Leader*	*Status Jan. 13*	*Status Feb. 13*
Compressed air leaks	Jerry Martin	Tags arrived	Identified (9) first level (25) second level leaks
Upgrade lighting	Don Wagner	Installed 22 lights in machine section	14 lights on order for casting

5.2 Project Methodology

As energy manager, you may run some projects, but many energy projects will be accomplished by others. Improvements can be implemented in many ways. You need to find the right one for the particular project and the company culture. Some methods are listed below.

1. *5 S.* Organizing the workplace to improve overall efficiency. The work is normally done as a team. The phases are

   Sort—keep essential items and store or throw away anything else.

   Set—identify storage for frequently used equipment close to where it is used.

   Shine—clean the workplace.

   Standardize—all areas and shifts should use the same methods.

   Sustain—maintain and audit practices.

   Pros: Changes are quickly implemented.

   Cons: Limited energy uses. Energy savings are harder to identify.

   Example project: Standardize and store furnace tools such as charging and skimming tools.

2. *Treasure hunts.* Method to identify and analyze potential energy savings. A team, both internal and external to the department, is used to walk through the plant and identify potential savings. This is normally done on a down-day to document idling practices. Another one to two days is then needed to observe practices when the line is running and to analyze potential savings. Treasure hunts can identify needs for a later Kaizen Blitz. More information is available at www.energystar.gov/treasurehunt.
   Pros: Quick identification of low-hanging fruit.
   Cons: Not in-depth. Problems are only identified, not fixed.
   Example: Identify lights that need motion sensors or that can be turned off on weekends. Shut off compressed air lines during lunch breaks.
3. *Kaizen* (or Kaizen Blitz or Kaizen Event). This activity uses certain tools to improve the process. Typically, a team meets for one week to brainstorm, trial, and implement improvements. In particular, waste such as energy waste can be identified and reduced or eliminated. Savings can be \$10,000–\$100,000 per year. Changes are normally process related, not capital related.
   Pros: Projects are implemented quickly. Using a team improves buy-in.
   Cons: Changes are incremental, not step changes.
   Example project: Better idling practices are identified and implemented. Burner control settings are optimized.
4. *Six Sigma.* This is a set of tools used for process improvement first developed by Motorola in 1985. A team is

formed with a trained Green Belt or Black Belt leader. The steps Define, Measure, Analyze, Implement, and Control are used. Statistical tools are used to analyze the data. Projects will typically last six months to one year. Given the amount of time required, savings of $250,000–$1,000,000 per year are needed, but rules vary per company. Changes are normally process related, not capital related.

Pros: Projects are large cost savings. Using a team improves buy-in.

Cons: The whole project, especially the Measure phase, can take a long time and requires company buy-in.

Example project: Reduction of hold time based on scheduling changes. Reduce off-analysis material.

5. *Energy value stream mapping (EVSM).* Analyzing and recording the flow of materials including temperatures and energy use. This can be done as a team or by one person. Savings are harder to identify but can come from analyzing the results.

   Pros: Identifies opportunities without major costs or time.

   Cons: No changes made, just identification of opportunities.

   Example project: Identify heat loss between processes.

6. *Preventative maintenance (PM).* Documented standards for maintenance. Specify time periods, steps, and settings.

   Pros: Required to maintain improvements.

   Cons: Need to make sure PM's aren't "pencil-whipped" (items checked off without being completed).

   Example: Burner tuning every three months.

7. *"Just-do-it" projects.* Smaller projects can be done without a team. Just do it! No capital is required. If money is needed, it could involve money from operations or maintenance accounts. Engineers, operations, or maintenance people could be used for these projects.

   Pros: Quick.

Cons: Need to make sure there is some type of control so the project gains aren't lost.

Example: Change furnace door open times by adjusting a valve.

8. *Engineering projects.* Project run by one or two engineers. Typically capital or major expense, but could be process changes.

   Pros: Required for many capital projects. These projects don't require a team.

   Cons: Engineering time and capital money can be difficult to get approval. Projects typically take longer given the capital request systems.

   Example: New burners or controls.

9. *Production preparation process (3P).* Method to eliminate waste through product and process design. Internal team meets for three to five days to brainstorm and analyze potential new equipment or process.

   Pros: Step-change with new equipment or new process.

   Cons: Expensive in time and normally in capital.

   Example: Elimination of steps in the painting room along with new energy-efficient equipment. Identification of preferred melt furnace and auxiliary equipment.

10. *Retro-commissioning (RCx).* Process to identify and improve inefficient equipment or systems. It is used to identify changes or resolves problems that have occurred over time. Normally done with a building, but may be used for equipment. Team may be internal or a mixture of internal and external members. Engineers, maintenance, and production are normally involved.

    Pros: Step-change to return the operation to original or better condition.

    Cons: Implementing all the recommendations can take a long time and can be expensive.

    Example: Team used training, maintenance, and proper tooling to bring performance of a melt furnace to designed energy intensity.

Chapter 6

Control and Evaluate Progress

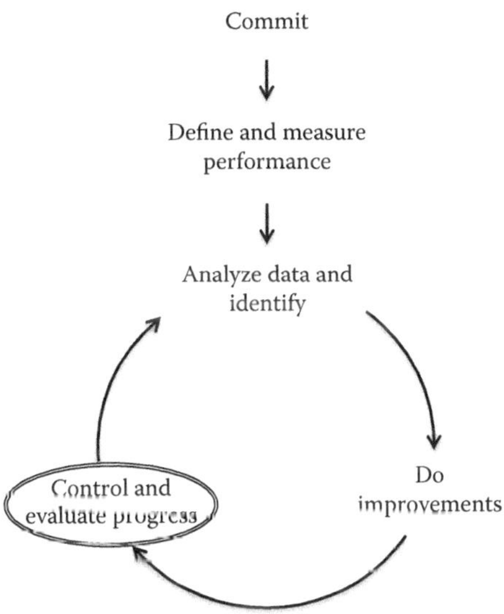

Doing all this work means nothing if you can't monitor and control the process. Remember that communication is important to publicize successes and spread knowledge. Achievements should be recognized! Certification can help in formalizing programs and to highlight your program to customers.

6.1 Track Progress

Establishing a tracking system is critical to understanding savings from projects, making sure your improvements continue, ensure there is no backsliding in performance in other areas, and identifying the next project. Make sure you track Key Performance Indicators (KPIs) at company, plant, and at the process/furnace level. Energy KPIs should not only be monitored by the energy manager, but should also be reported at the top level of management to keep focus on energy performance.

It is critical to track energy performance at least monthly through Excel® files and/or scorecards. These scorecards can be used to communicate effectively across the organization and to motivate supervisors.

Use spreadsheets and real-time analysis to analyze continuing performance.

1. Analyze energy efficiency achievements based on your established performance metrics. Watch for backsliders!
2. Analyze and evaluate. Watch for patterns and trends.
3. Look for both problems and improvements. Has performance slipped? Has some change improved energy usage? For instance, after a preventative maintenance (PM) made a major improvement on a furnace it was realized that the PM time period needed to decrease from one year to three months.
4. Complex, expensive software is NOT required. The file can be a simple Excel® spreadsheet. The program should be easy to maintain and update.
5. Short reports or emails can be sent based on your analysis. Even if supervisors have access to scorecards, they spend most of their time worrying about production, not worrying about energy. Notifying them of problems that are developing can help them, especially because furnace

energy problems may also affect furnace throughput. If possible, give them specific suggestions and recommendations. Let them know how much money is involved so they can set priorities. An example: "Gas use is rising as winter approaches. Heaters should be checked for the season that they are properly functioning and set to the right temperature setpoint. Check that the plant is ready for winter with doors closed, walls repaired, and summer fans shut off" or "Heat Treat Furnace #1 has been performing poorly since June. Maintenance may be required. We would lose roughly $8,000 per year if the current problem continues."

6. Longer reports showing equipment status and potential energy projects are appropriate. Send these to plant managers and other key personnel every three or six months.
7. Monitoring is needed, but work to help improve and not to blame people. It's a fine line.
8. Benchmarks should change every year or two. Remember, benchmarks are set to the first quartile. As improvements occur, the first quartile changes. Your goals should change to be more of a challenge.

POTENTIAL PITFALL!

Real-time or batch levels may show too much variation. Watch for trends over weeks or months.

6.2 Communication

6.2.1 Team Meetings

Make sure your team stays active. The team should meet every one to two months. Meeting in-person is best, then video conferencing, but phone/Internet conference calls work too.

Meet in person whenever possible. Hold meetings at different plants. Perhaps host an annual Energy Summit with everyone involved and presenting a topic.

Share best practices and transfer the information to other departments or plants. Document and share saving opportunities and best practices Transfer knowledge of successful (and unsuccessful) projects to other processes and plants. Don't keep reinventing the wheel!

Try to have each member speak during each meeting or schedule short presentations by certain team members. However, make sure every member has a chance to talk at every meeting to keep people involved. Review energy performance and project list status at the energy meetings. A shame factor works too. No one wants to present to his or her colleagues that no work was done since the last meeting.

Gather recommendations from the energy team and other key personnel on new projects and how to best run the energy program.

6.2.2 Energy Website

An *energy website* is a great method to document and share information. Consider what you want to include and who should have access. This could be for your team only, for management, everyone in the company, or even available to the public. Things you may want to include

1. Energy scorecards
2. Project status
3. Best practices, specifications, guidelines
4. Energy meeting minutes, newsletters
5. Online training
6. Event calendar

Documented best practices, standards, specifications, and guidelines can really help. Provide a general description of the change, potential savings, what processes are involved, details on how to implement the change, potential costs, and a contact for asking questions. These best practices can include things like how to adjust a burner, standard lighting for plants and offices, preventative maintenance for regenerative beds, or anything that makes sense for your company.

6.2.3 Grow Your Team

Besides your official team, everyone in your company can get involved and help. These are some things to try:

1. Educate staff on what they can do and the savings that will result.
2. Use posters on bulletin boards or in break rooms.
3. Hold contest for ideas to raise awareness and keep energy program fresh—actually implement winning ideas.
4. Get buy-in from management and all organizational areas affected by the action plan before finalizing it. No one likes to be blindsided by a change, even a good one.
5. Brainstorm with various departments to identify ways they can contribute. For example: purchasing could make sure new motors are NEMA Premium Efficiency while engineering should make sure they investigate state-of-the-art equipment versus "same old" when buying new equipment.
6. Work with vendors and customers to improve their energy management or to learn from them.
7. Try newsletters, emails, and open meetings. Include "At home" and "At work" energy saving tips.

   Example:

   "Energy Fact: Each CFL (compact fluorescent light) bulb you install in your house saves about $6 a year in electricity costs and can save more than $40 over its lifetime. At around $2 each, they quickly pay off the investment."

Monitor and communicate to everyone!

6.3 Recognize Achievements

Why should you recognize achievements?

1. It is a proven step for sustaining momentum and support for your energy program. Praise can go a long way.
2. Share credit when it is due to motivate staff and employees. You should reward and recognize teams or individuals with the best energy and cost saving ideas. This means people will be more likely to help the next time.
3. It validates the importance of your energy program.
4. It brings positive exposure to your energy program.
5. Establish an energy culture.
6. Build on successes and lessons learned.

Publicize successes internally to managers, employees, and stakeholders. Tailor your message for the audience. Internal recognition could include the following:

1. Newsletters, emails, intranet energy site, posters, signs
2. T-shirts, caps, hard hat stickers
3. CFL distribution to all employees
4. Safety meetings, energy meetings, Department of Energy (DOE), vendors
5. Award ceremonies
6. Employee bonus based on energy performance

Receive external recognition. Seek external recognition for your company from the U.S. Environmental Protection Agency (EPA) and others for achieving milestones such as becoming an ENERGY STAR Partner or DOE Better Plants Partner. Awards can be publicized on the company website; award logos and certificates can be displayed in the office waiting room, plant signage, or company trucks.

6.4 Certification

My personal opinion is that an energy program should be implemented for several years before seeking certification. Certification does not provide more energy savings. Certification requires more time and money. However, certification is a way to prove to your customers that you have a working energy program. Obtaining certification means that management fully supports the program now and in future years. It does not make sense to pursue certification and then drop the program after a few years.

For some companies and circumstances, certification is required. In general, certification options depend on your country though some certification is international. Some of the certifications that are available in the United States and internationally include the following.

DOE Better Plant Partner and Challenge: https://betterbuildingssolutioncenter.energy.gov/better-plants

- The DOE has a Better Buildings, Better Plants Program and Challenge. To sign up as a partner, the plant signs a challenge to meet a goal. The typical goal is to reduce energy intensity by 25% over 10 years. The DOE provides support with technical assistance and national recognition to those that sign up for the challenge.

U.S. ENERGY STAR: www.energystar.gov

- ENERGY STAR certified products are connected to a specific product, for example, refrigerators, air conditioners, or computers. Products made directly by the materials industry are not included in this program.
- ENERGY STAR building certification currently pertains to a specific building. For us, this would mean an office.

- ENERGY STAR partner is fairly easy to do. There is no cost. According to the website, becoming a partner means that you agree to: "Measure, track, and benchmark energy performance; Develop and implement a plan to improve energy performance, adopting the ENERGY STAR strategy; Educate your staff and the public about your partnership and achievements with ENERGY STAR." Becoming a partner gives you access to more resources such as meetings and awards.
- ENERGY STAR plant certification currently pertains to a specific plant. At this time (2016), metal plants are not included in the program though the website mentions that aluminum casting may be available soon.

ISO 14001—Environmental Management Standard: www.iso.org
ISO 14001 sets out the criteria for an environmental management system. While this is an environmental standard, it includes energy management. It does not state requirements for environmental performance, but maps out a framework that a company or organization can follow to set up an effective environmental management system. It can be used by any organization regardless of its activity or sector. Using ISO 14001:2004 can provide assurance to company management and employees as well as external stakeholders that environmental impact is being measured and improved.

The benefits of using ISO 14001 can include

- Reduced cost of waste management
- Savings in consumption of energy and materials
- Lower distribution costs
- Improved corporate image among regulators, customers, and the public

ISO 50001 Energy Management Standard Certification: www.iso.org
This can be helpful but can be costly for a company to implement. Certification may take two to four years. Most

companies explore certification based on customer requirements or to stand out from the competition.

ISO 50001 is a voluntary International Standard developed by the International Organization for Standardization (ISO) to provide organizations an internationally recognized framework to manage and improve their energy performance. It is similar to ISO 9001 and 14001. The standard addresses the following:

- Energy use and consumption
- Measurement, documentation, and reporting of energy use and consumption
- Design and procurement practices for energy-using equipment, systems, and processes
- Development of an energy management plan and other factors affecting energy performance that can be monitored and influenced by the organization

ISO 50001 does not include prescriptive energy performance goals. Rather, it provides a framework through which each organization can set and pursue its own goals for improving energy performance. It is possible to follow the standard but not go to the expense of certification.

Superior Energy Performance

The United States DOE's SEP certification requires the company to implement ISO 50001 and to pursue third-party verification of performance improvements. A study found that the average cost per facility to implement ISO 5001 and SEP certification was $319,000 per facility. This included internal labor along with external help and fees (Therkelsen et al. 2013). The payback of the program is expected to be less than two years.

6.5 Conclusion

Money and energy can be saved in the materials industry by combining the knowledge of energy management with the specialized knowledge of process heating. Use the steps of energy management along with the ideas in this book to help develop your energy program.

Although there are many options listed within this book on how to run an energy management program, you need to pick out the points that will work best for you given the resources you have. The only two requirements for an energy program are to utilize real data to analyze and monitor performance and to improve efficiency continuously with various energy projects Energy management programs should constantly evolve over time. Keep in mind that your business changes, industry changes, energy costs change, and technology changes. *Adapt!*

Resources are listed in Appendices I and II. There are many ideas for energy projects listed within this book and within Appendices III and IV.

As you run your energy program, keep in mind the potential pitfalls listed in this book. Running a program requires persistence, constant learning, and a sense of humor. Remember that saving energy helps the planet and helps your company financially at the same time.

And, most importantly, have fun!

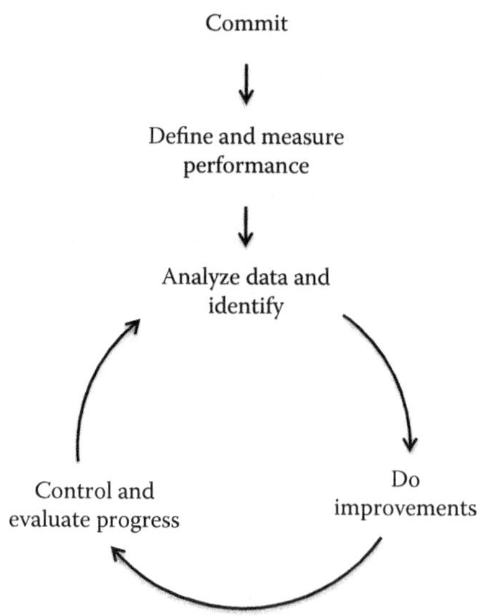

Appendix I: General Energy Resources

Resources for Internal Assessments

1. Utilize experts from other locations in your company (process heating, compressed air, etc.)
2. American Society of Mechanical Engineers (ASME) guidelines on system assessments—Process heating, compressed air, steam, pumps: www.asme.org
3. U.S. Department of Energy (DOE) Office of Energy Efficiency and Renewable Energy (EERE) systems training—Same systems as above + fans and insulation: www.energy.gov/eere/amo/training
4. Rutgers workbook for self-assessments: iac.university/technicalDocs/selfassessment.pdf
5. Use the Internet, books, and so on

Resources for External Assessments

1. DOE Industrial Assessment Centers (IACs; small and medium plants) and Industrial Technology Program Plant Assessments (large plants): energy.gov/eere/amo/industrial-assessment-centers-iacs.
2. Utilities—Explore financial and technical assistance offerings.

3. Vendors—May offer free system and equipment evaluations.
4. ENERGY STAR Partner and/or DOE Better Plants Partner—Industry peers and other manufacturers.
5. ESCOs (Energy Service Company)—An energy service company is a commercial business providing a range of comprehensive energy solutions including design, implementation of energy savings projects, energy conservation, energy infrastructure outsourcing, power generation and energy supply, and risk management. Explore energy performance contracting—no upfront capital!
6. Hire outside experts or consultants.

General Resources

1. Financial help—list of tax credits, grants, loans, rebates; Database of State Incentives for Renewables & Efficiency (DSIRE): www.dsireusa.org/
2. U.S. Department of Energy Advanced Manufacturing Office (DOE AMO): energy.gov/eere/amo/advanced-manufacturing-office
   - Publications, webinars, workshops: http://energy.gov/eere/amo/information-resources
3. U.S. EPA ENERGY STAR—energy management, recognition: www.energystar.gov

Appendix II: Materials Specific Resources

1. The Minerals, Metals, and Materials Society (TMS)—training and latest industry practices: www.tms.org
2. Papers
   a. Robert Eppich and Robert Naranjo, "Implementation of Metal Casting Best Practices," ITP Metal Casting, 2007: www1.eere.energy.gov/manufacturing/resources/metal casting/pdfs/implementation_final.pdf
   b. J. F. Schifo and J. T. Radia, "Theoretical/Best Practice Energy Use in Metalcasting Operations," May 2004: energy.gov/sites/prod/files/2013/11/f4/doebestpractice_052804.pdf
   c. BCS Incorporated, "Advanced Melting Technologies: Energy Saving Concepts and Opportunities for the Metal Casting Industry," 2005: www1.eere.energy.gov/manufacturing/resources/metalcasting/pdfs/advanced meltingtechnologies.pdf
   d. International Energy Agency (IEA), "Tracking Industrial Energy Efficiency and CO_2 Emissions," 2007: www.iea.org/publications/freepublications/publication/tracking_emissions.pdf
   e. U.S. Department of Energy Advanced Manufacturing Office (AMO) and the Industrial Heating Equipment Association, *Improving Process Heating System*

Performance: A Sourcebook for Industry, 3rd ed., 2015: energy.gov/eere/amo/downloads/improving-process-heating-system-performance-sourcebook-industry-third-edition

f. United Nations Industrial Development Organization and Ministry of International Trade and Industry, "Output of a Seminar on Energy Conservation in Iron and Steel Industry," 1992: www.unido.org/fileadmin/import/userfiles/puffk/iron.pdf

3. Combustion training and books
   a. Charles E. Baukal, Jr., John Zink Company, *The John Zink Combustion Handbook*, 2001
   b. Bloom Engineering, "Bloom Engineering Combustion Training Seminar": www.bloomeng.com/service/regional-combustion-seminar
   c. Multiple classes, CEC Combustion Safety: www.combustionsafety.com/solutions/training/combustion-university/
   d. E Instruments International, *E Instruments Combustion Handbook*, 2009: www.e-inst.com/combustion/
   e. Fives North American, "Effective Combustion and Its Control": combustion.fivesgroup.com/service/evolution/training.html
   f. *North American Combustion Handbook*, Vol. I, 3rd ed., 2010; *North American Combustion Handbook*, Vol. II, 3rd ed., 1997 (Original 1952)
   g. Hauck Manufacturing Company, "Combustion Seminar," "e-Solutions": www.hauckburner.com/
   h. Multiple videos, Industrial Heating Equipment Association: www.ihea.org/
   i. Multiple courses: www.johnzink.com/training-classes/
4. Heat transfer training and books
   a. Charles E. Baukal, Jr., *Heat Transfer in Industrial Combustion*, CRC Press, 2000
   b. "Refractory Heat Loss," FurnXpert

  c. Massachusetts Institute of Technology, "Intermediate Heat and Mass Transfer," "Radiative Transfer": www.ocw.mit.edu
  d. Arthur Morris, Gordon Geiger, and H. Alan Fien, *Handbook on Material and Energy Balance Calculations in Material Processing*, 3rd ed., 2011
5. Furnace systems training and books
  a. American Foundry Society (AFS)—training in casting practices: www.afsinc.org
  b. Association for Iron and Steel Technology (AIST)
  c. *Cupola Handbook*, AFS, 2008
  d. AIST, "Steel Mill Combustion and Thermal Systems," "Modern Electric Furnace Steelmaking": www.aist.org/
  e. ASM International, *ASM Handbook*, Vol. 15: Casting, 2008
  f. Barrie Jenkins and Peter Mullinger, *Industrial and Process Furnaces: Principles, Design and Operation*, Butterworth-Heinemann, 2008
  g. Seminars, SECO/WARWICK: www.secowarwick.com/
  h. W. Trinks, M. H. Mawhinney, R. A. Shannon, R. J. Reed, and J. R. Garvey, *Industrial Furnaces*, John Wiley & Sons, 2003
6. Energy management training and books
  a. Multiple courses and books, AEE: aeeprograms.com
  b. Specific system energy training and software, U.S. Department of Energy Advanced Manufacturing Office (DOE AMO): energy.gov/eere/amo/advanced-manufacturing-office
  c. Standards and guidelines on energy assessments, American Society of Mechanical Engineers (ASME): www.asme.org
  d. Energy Institute, "Training in Energy Management Through Open Learning (TEMOL)": environmental-expert.com/training/training-in-energy-management-through-open-learning-temol-148296

   e. Multiple webinars, Energy Star: www.energystar.gov
   f. World Energy Engineering Congress (WEEC): www.energycongress.com
7. Sustainability training and books
   a. Multiple courses, Coursera: www.coursera.org (Some of these are free while others may have minimal cost.)
   b. Multiple courses, Massachusetts Institute of Technology: www.ocw.mit.edu (These are all free.)

Appendix III: 250+ Ways to Reduce Industrial Utility Costs

Here are 250+ ways to reduce industrial utility costs, some of which you may not have considered yet.

Blowers

1. Locate air intakes to obtain appropriate air quality and best efficiency.
2. Use smooth, well-rounded air inlet ducts or cones for air intakes.
3. Minimize blower inlet and outlet obstructions.
4. Clean screens and filters regularly.
5. Use backward-inclined blower wheel designs.
6. Minimize blower speed.
7. Consider using a two-speed motor and running at the low speed whenever possible.
8. Use low-slip or no-slip belts.
9. Check belt tension regularly.
10. Eliminate variable pitch pulleys.
11. Use variable speed drives for large variable blower loads.
12. Use energy-efficient motors for continuous or near-continuous operation.

13. Use properly sized ductwork with appropriate bends and transitions.
14. Eliminate ductwork leaks.
15. Turn blowers off when they are not needed.

Boilers

16. Preheat combustion air with waste heat (watch out for NO_x concentration increases).
17. Use variable speed drives on large boiler combustion air fans with variable flows.
18. Burn wastes if permitted.
19. Insulate exposed heated oil tanks.
20. Inspect burner couplings, burner linkages, and damper linkages.
21. Clean burners, nozzles, strainers, and so on.
22. Inspect oil heaters for proper oil temperature.
23. Replace continuous pilots with electric ignition.
24. Close burner air and/or stack dampers when the burner is off to minimize heat loss up the stack.
25. Improve oxygen trim control (e.g., limit excess air to less than 10% on clean fuels).
26. Automate/optimize boiler blow-down. Recover boiler blow-down heat.
27. Use boiler blow-down to help warm the backup boiler.
28. Optimize deaerator venting.
29. Inspect door gaskets.
30. Inspect for scale and sediment on the water side.
31. Inspect for soot, fly ash, and slag on the fire side.
32. Optimize boiler water treatment.
33. Add an economizer to preheat boiler feed water using exhaust heat.
34. Isolate idling boilers to minimize losses due to induced flow of air.
35. Recycle steam condensate.
36. Avoid oversizing—match the connected load.

37. Study part-load characteristics and cycling costs to determine the most efficient mode for operating multiple boilers.
38. Consider multiple modular boiler units instead of one or two large boilers.
39. Use a small boiler for a small summer load.
40. Establish a boiler efficiency-maintenance program.

Buildings

41. Seal exterior cracks/openings/gaps with caulk, gaskets, weather stripping, and so on.
42. Consider new thermal doors, thermal windows, roofing insulation, and so on.
43. Install windbreaks near exterior doors.
44. Replace single-pane glass with insulating glass.
45. Consider covering some window and skylight areas with insulated wall panels inside the building.
46. If visibility is not required but light is required, consider replacing exterior windows with insulated glass block.
47. Consider tinted glass, reflective glass, coatings, awnings, overhangs, draperies, blinds, and shades for sunlit exterior windows.
48. Use landscaping to advantage.
49. Add vestibules or revolving doors to primary exterior personnel doors.
50. Consider automatic doors, air curtains, strip doors, and so on at high-traffic passages between conditioned and nonconditioned spaces. Use self-closing doors if possible.
51. Use intermediate doors in stairways and vertical passages to minimize building stack effect.
52. Use dock seals at shipping and receiving doors.
53. Bring cleaning personnel in during the working day or as soon after as possible to minimize lighting and HVAC costs.
54. Utilize reflective roofing materials.

Chillers

55. Increase the chilled water temperature set point if possible.
56. Use the lowest temperature condenser water available that the chiller can handle.
57. Clean heat exchangers when fouled.
58. Optimize condenser water flow rate and refrigerated water flow rate.
59. Replace old chillers or compressors with new higher-efficiency models.
60. Use water-cooled rather than air-cooled chiller condensers.
61. Use energy-efficient motors for continuous or near-continuous operation.
62. Specify appropriate fouling factors for condensers.
63. Use the most efficient refrigerant.
64. Do not overcharge refrigerant.
65. Do not overcharge oil.
66. Install a control system to coordinate multiple chillers.
67. Study part-load characteristics and cycling costs to determine the most efficient mode for operating multiple chillers.
68. Run the chillers with the lowest operating costs to serve base load.
69. Avoid oversizing—match the connected load.
70. Isolate off-line chillers and cooling towers.
71. Establish a chiller efficiency-maintenance program.

Compressed Air

72. Install a control system to coordinate multiple air compressors.
73. Study part-load characteristics and cycling costs to determine the most efficient mode for operating multiple air compressors.

74. Avoid oversizing—match the connected load.
75. Load up modulation-controlled air compressors (they use almost as much power at partial load as at full load).
76. Turn off the backup air compressor until it is needed.
77. Reduce air compressor discharge pressure to the lowest acceptable setting (this will reduce compression energy requirements and compressed air leakage rates).
78. Find the operation that requires the highest pressure and work to reduce the need (e.g., pressure drops in piping).
79. Use the highest reasonable dryer dew point settings.
80. Turn off refrigerated and heated air dryers when the air compressors are off.
81. Use a control system to minimize heatless desiccant dryer purging.
82. Minimize purges, leaks, excessive pressure drops, and condensation accumulation.
83. Use solenoids to shut off air lines when not needed (shut off blow-offs and potential leaks).
84. Use drain controls instead of continuous air bleeds through the drains.
85. Consider engine-driven or steam-driven air compression to reduce electrical demand charges.
86. Change the electric motors to an energy-efficient design.
87. Replace standard v-belts with high-efficiency cogged v-belts as the old v-belts wear out.
88. Use a small air compressor for HVAC load when production load is off.
89. Take air compressor intake air from the coolest (but not air conditioned) location.
90. Use an air-cooled aftercooler to heat building makeup air in winter.
91. Be sure that heat exchangers are not fouled (e.g., with oil).
92. Be sure that air/oil separators are not fouled.
93. Monitor pressure drops across suction and discharge filters and clean or replace filters promptly upon alarm.
94. Use a properly sized compressed air storage receiver.

95. Minimize disposal costs by using lubricant that is fully demulsible and an effective oil–water separator.
96. Consider alternatives to compressed air such as blowers for cooling, hydraulic rather than air cylinders, electric rather than air actuators, and electronic rather than pneumatic controls.
97. Use nozzles or venturi-type devices rather than blowing with open compressed air lines.
98. Check for leaking drain valves on compressed air filter/regulator sets. Certain rubber-type valves may leak continuously after they age and crack.
99. In dusty environments, control packaging lines with high-intensity photocell units instead of standard units with continuous air purging of lenses and reflectors.
100. Establish a compressed air efficiency-maintenance program.

Compressors

101. Consider variable speed drive for variable load on positive displacement compressors.
102. Use a synthetic lubricant if the compressor manufacturer permits it.
103. Be sure lubricating oil temperature is not too high (oil degradation and lowered viscosity) and not too low (condensation contamination).
104. Change the oil filter regularly.
105. Periodically inspect compressor intercoolers for proper functioning.
106. Use waste heat from a very large compressor to power an absorption chiller or preheat process or utility feeds.
107. Establish a compressor efficiency-maintenance program.

Condensate

108. Return steam condensate to boiler feedwater where the return system cost is justified.

109. Inspect steam traps regularly.
110. Use HVAC condensate to avoid makeup water and sewer costs (e.g., use HVAC condensate for cooling tower makeup water).

Cooling Towers

111. Control cooling tower fans based on leaving water temperatures.
112. Control to the optimum water temperature as determined from cooling tower and chiller performance data.
113. Use two-speed or variable speed drives for cooling tower fan control if the fans are few. Stage the cooling tower fans with on–off control if there are many.
114. Turn off unnecessary cooling tower fans when loads are reduced.
115. Cover hot water basins (to minimize algae growth that contributes to fouling).
116. Balance flow to cooling tower hot water basins.
117. Periodically clean plugged cooling tower water distribution nozzles.
118. Install new nozzles to obtain a more uniform water pattern.
119. Replace splash bars with self-extinguishing polyvinyl chloride (PVC) cellular-film fill.
120. On old counterflow cooling towers, replace old spray-type nozzles with new square-spray ABS practically nonclogging nozzles.
121. Replace slat-type drift eliminators with high-efficiency, low-pressure-drop, self-extinguishing.
122. Replace wood slats with PVC cellular units.
123. If possible, follow the manufacturer's recommended clearances around cooling towers and relocate or modify structures, signs, fences, dumpsters, and so on that interfere with air intake or exhaust.
124. Optimize cooling tower fan blade angle on a seasonal and/or load basis.

125. Correct excessive and/or uneven fan blade tip clearance and poor fan balance.
126. Use a velocity pressure recovery fan ring.
127. Divert clean air-conditioned building exhaust to the cooling tower during hot weather.
128. Reline leaking cooling tower cold water basins.
129. Check water overflow pipes for proper operating level.
130. Optimize chemical use.
131. Consider side stream water treatment.
132. Restrict flows through large loads to design values.
133. Shut off loads that are not in service.
134. Locate make-up and blow-down tie-ins to minimize heat tracing cost in winter.
135. Take blow-down water from the return water header.
136. Optimize blow-down flow rate.
137. Automate blow-down to minimize it.
138. Send blow-down to other uses or to the cheapest available sewer allowed. (Remember, the blow-down does not have to be removed at the cooling tower. It can be removed anywhere in the piping system.)
139. Implement a cooling tower winterization plan to minimize ice buildup.
140. Avoid simultaneous water heating and cooling in cold weather.
141. Install interlocks to prevent fan operation when there is no water flow.
142. Establish a cooling tower efficiency-maintenance program.

Distillation

143. Optimize it. If the distillation process is old, the utility costs have probably changed significantly, so reoptimize it.
144. Use distillation bottoms heat to preheat distillation feeds.

Drives

145. Use variable speed drives for large variable loads. (Remember that engines can be variable-speed too.)
146. Use high-efficiency gear sets.
147. Use precision alignment.
148. Check belt tension regularly.
149. Eliminate variable pitch pulleys.
150. Use synchronous belts as no-slip alternatives to v-belts.
151. Use synthetic lubricants for large gearboxes.
152. Eliminate eddy current couplings.
153. When they're not needed, shut them off.

Drying

154. Reduce moisture content of feed material to the maximum extent possible by mechanical methods.
155. Operate dryers only when necessary.
156. Consider alternative technologies (e.g., infrared, microwave).
157. Minimize dryer heat loss (e.g., insulate, seal, recirculate).
158. Recover heat from dryer outlet streams.

Electricity

159. Shop for the best price in the deregulated electricity market.
160. Know your historical demand profile so you can make the best deal in the deregulated electricity market.
161. If your load factor is poor, aggregate with other sites that have higher load factors so you can make the best deal in the deregulated electricity market.
162. Shift loads to off-peak times.
163. Know the cost of setting a higher monthly peak demand and set alarms and shut down equipment accordingly.

164. Stagger startup times for equipment with large starting currents to minimize load peaking.
165. Use standby electric generation equipment for on-peak high-load periods.
166. Test emergency generators during on-peak high-load periods.
167. Delay startup of new equipment to prevent a new summer peak.
168. If possible, shut off a piece of equipment before starting the alternate piece.
169. Obtain a sales tax exemption if available (available for electricity used for manufacturing in many states).
170. Consider alternative electric rate schedules (e.g., interruptible rate).
171. Take advantage of utility rebate programs (e.g., lighting upgrade, thermal storage, energy efficient motors).
172. Correct power factor to at least 90%.
173. Buy the substation, transformer, and so on from the utility.
174. Relocate transformers close to main loads.
175. Set transformer taps to optimum settings.
176. Disconnect primary power to transformers that do not serve any active loads (e.g., seasonal loads or surplus transformers).
177. Consider on-site electric generation or cogeneration.
178. Have the utility "buy out" your cogeneration plan (i.e., reduce your rate if you will terminate the project).
179. Check electric meter printouts and electric bills for use during outages and holiday shut-down periods.
180. Shut off unnecessary computers, printers, and copiers at night.
181. Buy ENERGY STAR compliant products.
182. Optimize motor selection using free DOE Motor Challenge Program software (Motor Master).

Engines, Stationary

183. Use waste heat to power an absorption chiller or preheat process or utility feeds.
184. Do not pay motor vehicle tax on oil for heating equipment or stationary engines.

Fans

185. Use smooth, well-rounded air inlet cones for fan air intakes.
186. Avoid poor flow distribution at the fan inlet.
187. Minimize fan inlet and outlet obstructions.
188. Clean screens, filters, and fan blades regularly.
189. Use airfoil-shaped fan blades.
190. Minimize fan speed.
191. Consider using a two-speed motor and running at the low speed whenever possible.
192. Use low-slip or no-slip belts.
193. Check belt tension regularly.
194. Eliminate variable pitch pulleys.
195. Use variable speed drives for large variable fan loads.
196. Use energy-efficient motors for continuous or near-continuous operation.
197. Use properly sized ductwork with appropriate bends and transitions.
198. Eliminate ductwork leaks.
199. Turn fans off when they are not needed.
200. Properly size fans for the system needs.

Fire Protection Systems

201. Test electric fire pumps during electrical off-peak periods.
202. Automate pressure maintenance pumps to run only when water pressure is low.

203. Check for continuous overflow at storage tank due to an open feed.
204. Check water overflow pipes for proper operating level.
205. Consider using the storage tank for refrigerated water thermal storage.

Heat Exchangers

206. Specify appropriate heat exchanger fouling factors.
207. Maintain “self-cleaning” fluid velocities.
208. Clean heat exchangers when excessively fouled.
209. Turn off the flow when the heat exchanger is out of service.
210. Routinely log process data for assessment of fouling and establish a heat exchanger efficiency-maintenance program.

HVAC (Heating/Ventilation/Air Conditioning)

211. Tune up the HVAC control system.
212. Consider installing a building automation system (BAS) or energy management system (EMS) or restoring an out-of-service one.
213. Balance the system to minimize flows and reduce blower/fan/pump power requirements.
214. Eliminate or reduce reheat whenever possible.
215. Prevent unauthorized thermostat adjustments.
216. Use appropriate HVAC thermostat setback.
217. Raise “cool” settings and lower “heat” settings to appropriate levels.
218. Use morning precooling in summer and preheating in winter (i.e., before electrical peak hours).
219. Use building thermal lag to minimize HVAC equipment operating time.
220. In winter during unoccupied periods, allow temperatures to fall as low as possible without freezing water lines or damaging stored materials.

221. In summer during unoccupied periods, allow temperatures to rise as high as possible without damaging stored materials.
222. Improve control and utilization of outside air.
223. Use air-to-air heat exchangers to reduce energy requirements for heating and cooling of outside air.
224. Reduce HVAC system operating hours (e.g., nights, weekends).
225. Optimize ventilation.
226. Ventilate only when necessary. To allow some areas to be shut down when unoccupied, install dedicated HVAC systems on continuous loads (e.g., computer rooms).
227. Provide dedicated outside air supply to kitchens, cleaning rooms, combustion equipment, and so on to avoid excessive exhausting of conditioned air.
228. Use evaporative cooling in dry climates.
229. Reduce humidification or dehumidification during unoccupied periods.
230. Use atomization rather than steam for humidification where possible.
231. Clean HVAC unit coils periodically and comb mashed fins.
232. Upgrade filter banks to reduce pressure drop and thus lower fan power requirements.
233. Check HVAC filters on a schedule (at least monthly) and clean/change if appropriate.
234. Check pneumatic controls air compressors for proper operation, cycling, and maintenance.
235. Isolate air conditioned loading dock areas and cool storage areas using high-speed doors or clear PVC strip curtains.
236. Install ceiling fans to minimize thermal stratification in high-bay areas.
237. Relocate air diffusers to optimum heights in areas with high ceilings.
238. Consider reducing ceiling heights.

239. Eliminate obstructions in front of radiators, baseboard heaters, and so on.
240. Check reflectors on infrared heaters for cleanliness and proper beam direction.
241. Use professionally designed industrial ventilation hoods for dust and vapor control.
242. Use local infrared heat for personnel rather than heating the entire area.
243. Use spot cooling and heating (e.g., use ceiling fans for personnel rather than cooling the entire area).
244. Purchase only high-efficiency models for HVAC window units.
245. Put HVAC window units on timer control.
246. Improve crude temperature control of multiple HVAC window units.
247. Control infrared heaters based on ambient temperature.
248. Don't oversize cooling units (oversized units will "short cycle," which results in poor humidity control).
249. Use a system of multiple modular boilers instead of one large boiler.
250. Install multifueling capability and run with the cheapest fuel available at the time.
251. Consider dedicated makeup air for exhaust hoods. (Why exhaust the air conditioning or heat if you don't need to?)
252. Minimize HVAC fan speeds.
253. Consider desiccant drying of outside air to reduce cooling requirements in humid climates.
254. Consider ground source heat pumps.
255. Seal leaky HVAC ductwork.
256. Seal all leaks around coils.
257. Repair loose or damaged flexible connections (including those under air handling units). (The harder they are to get to, the more likely they are to be loose since installation!)
258. Eliminate simultaneous heating and cooling during seasonal transition periods.

259. Consider directing clean conditioned exhaust air to the inlet of a cooling tower, air-cooled condenser, or evaporative cooler when doing so will reduce energy requirements.
260. Zone HVAC air and water systems to minimize energy use.
261. Inspect, clean, lubricate, and adjust damper blades and linkages.
262. Establish an HVAC efficiency-maintenance program.
263. Install low-emittance windows or window coatings.
264. Use waste heat from air compressors or furnaces for plant heat.
265. Reduce the heat load (ex. lights, copiers, computers left on overnight).
266. Use solar air heating.

Insulation

267. Repair damaged insulation.
268. Insulate any hot or cold metal or insulation.
269. Replace wet insulation.
270. Use an infrared gun to check for cold wall areas during cold weather or hot wall areas during hot weather.
271. Consider adding insulation to water heaters, etc. located in unheated/uncooled spaces.
272. Insulate HVAC ducts running outside and through unoccupied spaces.
273. Take vapor barriers and insulation coverings very seriously.

Lighting

274. Reduce excessive illumination levels to standard levels using switching delamping, and so on. (Know the electrical and lumen effects before doing delamping.)

275. Aggressively control lighting with clock timers, delay timers, photocells, and/or occupancy sensors.
276. Install efficient alternatives to incandescent lighting, mercury vapor lighting, and so on.
277. Select ballasts and lamps carefully with a high power factor and long-term efficiency in mind.
278. Upgrade obsolete fluorescent systems to T-8 lamps and electronic ballasts.
279. Consider T-5 and/or LED lighting systems.
280. Consider lowering the fixtures to enable using fewer fixtures of them.
281. Consider painting the walls a lighter color and using less lighting fixtures or lower wattages.
282. Use task lighting and reduce background illumination.
283. Reevaluate exterior lighting strategy, type, and control—control it aggressively.
284. Change exit signs from incandescent to LED.
285. Use daylighting.

Miscellaneous

286. Meter any unmetered utilities. Know what is normal efficient use. Track down causes of deviations.
287. Shut down spare, idling, or unneeded equipment.
288. Make sure that all of the utilities to "mothballed" areas are turned off, including utilities such as compressed air and cooling water.
289. Install automatic control to coordinate multiple air compressors, chillers, cooling tower cells, boilers, and so on efficiently.
290. Renegotiate utilities contracts to reflect current loads and variations.
291. Consider buying utilities from neighbors, particularly to handle peaks.

292. Leased space often has low-bid inefficient equipment. Consider upgrades if your lease will continue for five or more years.
293. Take as much of the sales tax exclusion on manufacturing utilities as you can take.
294. Avoid late payment charges.
295. Use alternative sampling systems to eliminate sample hoods, thus reducing water, sewer, and ventilation requirements.
296. Stop using rental equipment for long-term requirements. Install permanent high-efficiency equipment where appropriate.
297. Adjust fluid temperatures within acceptable limits to minimize undesirable heat transfer in long pipelines.
298. Minimize use of flow bypasses and minimize bypass flow rates.
299. Provide restriction orifices in purges (nitrogen, steam, etc.).
300. Eliminate unnecessary flow measurement orifices.
301. Consider alternatives to large pressure drops across valves.
302. Turn off winter heat tracing that is on in summer.

Motors

303. Properly size to the load for optimum efficiency.
304. Use energy-efficient motors where economical.
305. Use synchronous motors to improve power factor.
306. Check alignment.
307. Check for undervoltage and overvoltage conditions.
308. Balance the three-phase power supply.
309. Require efficiency restoration from motor rewinding using ANSI/EASA AR100-2015 or buy new.
310. Add automation to turn them off when not needed.
311. Use a systems approach to investigate and optimize motors, drives, and equipment (pumps, fans, etc.).
312. Use soft starters.

Natural Gas

313. Shop for the best natural gas price.
314. Minimize natural gas spot market purchases, especially in winter.
315. Bypass the local natural gas distributor if possible. Pipe directly to transmission if nearby (or just seriously threaten to and negotiate a lower rate).
316. Install a fuel oil or propane backup system and switch load from firm to interruptible service.
317. Consider replacing pilots with electric ignition devices.
318. Obtain a sales tax exemption if available (available for natural gas used for manufacturing in many states).
319. Consider natural gas for summer cooling (incentives, rebates, electrical demand reduction).

Nitrogen and Other Specialty Gases

320. Shop for the best specialty gas prices.
321. Generate nitrogen or oxygen with a membrane system where lower quality than cryogenic will suffice.
322. Use liquid nitrogen vaporization for cooling.
323. Use liquid nitrogen vaporization for vent stream volatile organic compounds (VOCs) condensation.
324. Periodically survey gas systems (especially oxygen, nitrogen, and other high cost gases) using ultrasonic and other leak detection methods and fix leaks.

Pumps

325. Operate pumping near the best efficiency point.
326. Modify pumping to minimize throttling and bypass loops.
327. Adapt to wide load variation with variable speed drives or sequenced control of smaller units.
328. Stop running both pumps—add an auto-start for an on-line spare or add a booster pump in a problem area.

329. Use booster pumps for small loads requiring higher pressures.
330. Increase fluid temperature differentials to reduce pumping rates.
331. Repair seals and packing to minimize water waste.
332. Balance the system to minimize flows and reduce pump power requirements.
333. Use siphon effect to advantage: don't waste pumping head with a free-fall return.
334. Shut off pumps when not needed.

Refrigeration

335. Use water-cooled condensers rather than air-cooled condensers.
336. Challenge the need for refrigeration, particularly for old batch processes.
337. Avoid oversizing—match the connected load.
338. Consider gas-powered refrigeration equipment to minimize electrical demand charges.
339. Use "free cooling" to allow chiller shutdown in cold weather.
340. Use refrigerated water loads in series if possible.
341. Convert firewater or other tanks to thermal storage.
342. Don't assume that the old way is still the best—particularly for energy-intensive low-temperature systems.
343. Correct inappropriate brine or glycol concentration that adversely affects heat transfer and/or pumping energy.
344. If it sweats, insulate it. If it is corroding, replace it first.
345. Consider adding hot gas bypass if it will reduce power consumption.
346. Make adjustments to minimize hot gas bypass operation.
347. Insulate hot gas lines that are inside buildings.
348. Inspect moisture/liquid indicators.
349. Consider change of refrigerant type if it will improve efficiency.

350. Check for correct refrigerant charge level.
351. Inspect the purge for air and water leaks.
352. Establish a refrigeration efficiency-maintenance program.

Steam

353. Fix steam leaks and condensate leaks.
354. Accumulate work orders for repair of steam leaks that can't be fixed during the heating season because of system shutdown requirements. Tag each such leak with a durable tag with a good description.
355. Use let-down steam turbines to produce lower steam pressures.
356. Use more efficient steam desuperheating methods.
357. Inspect steam traps regularly and repair malfunctioning traps promptly.
358. Consider recovery of vent steam (e.g., on large flash tanks).
359. Use waste steam for water heating.
360. Use an absorption chiller to condense exhaust steam before returning the condensate to the boiler.
361. Use electric pumps instead of steam eductors for sump clean-out, or at least provide reliable automatic shut-off.
362. Establish a steam efficiency-maintenance program.

Steam Turbines

363. Reevaluate options if efficiency is less than 65% (e.g., avoid small crude steam turbines).
364. Use steam turbines for large volume pressure letdown.
365. Use a piggy-back absorption system where existing chillers are driven by steam turbines.
366. Establish a steam turbine efficiency-maintenance program.

Vacuum Systems

367. Reduce excessive vacuum system air bleed-in (e.g., from unplanned leaks and from pressure controls).
368. Operate steam jets at design steam pressure (i.e., not too high or too low).
369. Avoid steam jets where pumps should be used (e.g., higher flows and pressures not requiring alloys above type 304 stainless steel).
370. Shut down steam jets that never shut down on batch units.
371. Modify oversized (overspecified) steam jet systems.
372. Repair worn steam jet parts (e.g., greater than 5% overage on the steam nozzle throat).
373. Replace third-stage ejectors with vacuum pumps.
374. Establish a vacuum system efficiency-maintenance program.

Waste Recovery

375. Recover heat from flue gas, engine cooling water, engine exhaust, low-pressure waste steam, drying oven exhaust, boiler blow-down, and so on.
376. Recover fuel from vent streams.
377. Recover heat from thermal oxidizer or incinerator off-gas.
378. Use waste heat for fuel oil heating, boiler feed water heating, outside air heating, and so on.
379. Use chiller waste heat to preheat hot water.
380. Use heat pumps.
381. Use absorption refrigeration.
382. Use thermal wheels, run-around systems, heat pipe systems, and air-to-air exchangers.

Water and Sewer

383. Recycle water, particularly for uses with less critical quality requirements.
384. Recycle water, especially if sewer costs are based on water consumption.
385. Balance closed systems to minimize flows and reduce pump power requirements.
386. Eliminate once-through cooling with water.
387. Use the least expensive type of water that will satisfy the requirement.
388. Fix water leaks.
389. Test for underground water leaks. (It's easy to do over a holiday shutdown.)
390. Check water overflow pipes for proper operating level.
391. Automate blow-down to minimize it.
392. Provide proper tools for wash down—especially self-closing nozzles.
393. Install efficient irrigation.
394. Reduce flows at water sampling stations.
395. Eliminate continuous overflow at water tanks.
396. Promptly repair leaking toilets and faucets.
397. Use water restrictors on faucets, showers, and so on.
398. Use self-closing type faucets in restrooms.
399. Use the lowest possible hot water temperature.
400. Do not use a heating system hot water boiler to provide service hot water during the cooling season—install a smaller, more efficient system for the cooling season service hot water.
401. If water must be heated electrically, consider accumulation in a large insulated storage tank to minimize heating at on-peak electric rates.
402. Use multiple, distributed, small water heaters to minimize thermal losses in large piping systems.
403. Use freeze protection valves rather than manual bleeding of lines.

404. Consider leased and mobile water treatment systems, especially for deionized water.
405. Seal sumps to prevent seepage inward from necessitating extra sump pump operation.
406. Install pretreatment to reduce total organic carbon (TOC) and biochemical oxygen demand (BOD) surcharges.
407. Verify the water meter works.
408. Verify the sewer flows if the sewer bills are based on them.

Yes, this is much more than 250 items. I did not originate this list but found it on the Internet approximately 20 years ago and have added to it. Now I can't find the original source. My apologies and thanks to the original author or authors!!!

Appendix IV: Materials Specific Energy Saving Ideas

Furnace Operations

1. Fully utilize furnaces.
2. Shut down extra furnaces and similar equipment.
3. Fully utilize the most energy-efficient furnaces.
4. Minimize charge time (charge tables, prep metal, etc.).
5. Standardize the furnace feed rate (stabilize energy usage).
6. Quick door open/close times.
7. Optimize skim time. Keep dross/slag levels low for better heat transfer. However, the molten surface tends to reflect heat and forms more oxidized material.
8. Minimize heat loss during alloy checks.
9. Reduce hold time for chemistry.
10. Reduce casting, hold, and unloading times.
11. Minimize door open time.
12. Reduce hold times for transportation (cranes, loaders, etc.).
13. Display furnace temp info for operators.

14. Optimize and standardize idle practices on furnaces including auxiliary equipment (baghouses, fans, etc.).
15. Optimize hot air pulled into the baghouses—do not pull more heat than necessary from the furnace.
16. Reduce heat lost to cooling water or to atmospheres such as nitrogen or argon.
17. Reduce the weight and/or specific heat of the rack or conveyor in heat treat furnaces.
18. Keep the rack or conveyor hot between uses.
19. Improve maintenance practices to reduce downtime and optimize performance.
20. Utilize the best planning system possible to optimize the system.
21. Include temperatures and heat loss within scheduling programs.

Metal

22. Charge clean scrap (little sand, rust, or other impurities).
23. Keep charges dry (under roof).
24. Use fine turning quickly and store indoors to reduce metal oxidation.
25. Improve raw material quality.
26. Purchase molten metal (aluminum) versus sows.
27. Work toward improving the alloying and charge mix process so charges are never off-specification.
28. Reduce heat loss between steps with better scheduling or covered transfer methods.

Manpower

29. Look for best practices from the best performing shift.
30. Stagger shifts and lunches.
31. Provide training such as combustion training.
32. Include plant workers on the energy team.

Burners and Combustion

33. Optimize the air/fuel ratio.
34. Use a gas analyzer to adjust the air/fuel ratio automatically.
35. Reduce high fire to the minimum required.
36. Reduce low fire to the minimum required.
37. Use the pilot relight instead of low fire.
38. Use hot air burners—regen or recuperative.
39. Use indoor air for combustion (slightly higher temperature than outside but must be clean).
40. Utilize pulse burners where appropriate.
41. Utilize oxy-fuel burners where appropriate.
42. Reduce compressed air humidity with air dryer/dehumidifier.
43. Maintain burners—clean metal splash or carbon build-up to keep air/fuel ratio correct.
44. Use chemical energy where appropriate.
45. Use a virtual lance booster (VLB) to increase the efficiency of chemical energy (steel).
46. Use computer modeling to optimize the fire rate and angle of burners.
47. Optimize burner sizes—oversized burners are less efficient.
48. Investigate newer type burners such as porous or self-regenerative.
49. Investigate alternate methods of ladle and crucible heating such as self-regenerative burners, radiation burners, or imbedded electrical heaters.

Furnace Settings

50. Reduce bath temperature setpoint to the minimum.
51. Cast at low temperatures to reduce energy for cooling.
52. Watch for and reduce temperature overshoot.

53. Reduce control temperature setpoints (roof, flue, etc.).
54. Use metal bath stirring and optimize the settings.
55. Maintain proper furnace pressure.

Refractory and Furnace Design

56. Perform regular inspections of refractory and steel to schedule repairs.
57. Improve insulation values to reduce heat loss while keeping the freeze plane in mind.
58. Repair all furnace door seals and furnace openings.
59. Build covers for wells and openings—mechanical opening is preferred.
60. Paint outside steel silver.
61. Reduce airflow on sides of furnaces (no fans).
62. Design furnaces to increase residence time.
63. Move openings in heat treat furnaces to the lowest position possible.
64. Add cleats and chains to increase dryer efficiency.

Waste Heat

65. Transport, retain heat, and charge hotter material from previous process.
66. Preheat combustion air—in descending order of efficiency—regenerative burners, recuperative burners, recuperative heat exchangers, low-temperature combustion air, inside air.
67. Preheat incoming material—zones, stacks, ledges, boxes, and so on.
68. Use heat for lower temperature operations (i.e., aging).
69. Use heat to heat water (office or operation water tanks).
70. Use heat for plant heat.
71. Use heat to make electricity.
72. Use combined heat and power (CHP) units.

73. Use waste heat from or reduce heat to ash, slag, dross, or other byproducts.

Electricity

74. Move work to lower-energy-cost time periods.
75. Install capacitor banks to lower power factor below the utility required value.
76. Reduce peak demand with automated controls.
77. Use variable frequency drives (VFDs) on exhaust blowers to minimize heat loss.
78. Use VFDs on fans for better control furnace temperatures.
79. Use VFDs on pumps, casters, air compressors, and other appropriate locations.
80. Install new ultra-high-power transformers to reduce energy losses and increase throughput (DOE AMO and IHEA 2015).
81. Clean and maintain reflectors, end caps, and emitters in electric infrared systems (DOE AMO and IHEA 2015).
82. In induction furnaces, use high-efficiency solid-state power supplies for less heat loss in the power supply (DOE AMO and IHEA 2015).
83. In induction furnaces, metal should be charged as densely as possible to ensure good electric coupling (ENERGY STAR 2016b).
84. In induction furnaces and electric arc furnaces, leave a liquid heel between batches (ENERGY STAR 2016b).
85. In induction furnaces, upgrade from a low-frequency to a high-frequency system (ENERGY STAR 2016b).
86. Investigate UV-based systems versus gas curing ovens and oxidizers for paint lines (DOE AMO and IHEA 2015).
87. Gate communicated turn-off (GCT) Inverters save energy in large variable speed AC drives to decrease switching losses (Worrell et al. 2010).

Process

88. Reduce product scrap rate.
89. Reduce planned and unplanned recovery loss (ends, alloy changes, gates, flash, etc.).
90. Reduce metal loss (oxidation).
91. Use alloy that reduces required heat treatment.
92. Utilize near net shape methods (better surface, less distortion, etc.).
93. Incorporate semisolid processing such as thixoforming or rheocasting.
94. Increase use of scrap material (scrap material has less embedded energy than prime material).
95. Utilize step saving processes such as continuous casting and direct reduction ironmaking.
96. Watch that heat treat furnaces are not overloaded. This changes circulation and reduces gas residence time.

Measurement

97. Use time in the process as a method to estimate energy usage.
98. Consider the positions of sensors carefully to receive more accurate data.
99. Add flowmeters to individual processes to optimize systems.
100. Use mass flow with pressure and temperature compensation for best control on burner systems.

Research

101. Work with universities and research laboratories on difficult problems.
102. Use software and calculations to trial changes before real-life trials.
103. Research new processes with energy as a design parameter.

Plant

104. Perform gas leak checks (sniffers, infrared camera, etc.).
105. Move air compressors into one area to optimize maintenance and improve control.
106. Move HVAC (Heating, Ventilating, and Air Conditioning) systems to one area to optimize maintenance and improve control.
107. In slurry pipelines, optimize flow—the highest flow may not be best.
108. Investigate the need for clean water like deionized water—use the lowest treatment required for the process.

References

AMC (American Metalcasting Consortium). 2016. Metalcasting Industry Roadmap.

Baldwin, S. 2015. Quadrennial Technology Review-2015. U.S. Department of Energy.

Belt, C. 2004. Energy Efficiency Tests in Aluminum Combination Melting and Holding Furnaces. *Light Metals 2004*, ed. A. Tabereaux, 613–617. Warrendale, PA: TMS.

Belt, C. 2012. Energy Opportunities in the Aluminum Processing Industry. *Energy Technology 2012: Carbon Dioxide Management and Other Technologies*, ed. M. Salazar-Villalpando, 57–64. Hoboken, NJ: John Wiley & Sons.

Belt, C. 2013. Overview of Industrial Energy Training and Software. *Energy Technology 2013*. ed. S. Pati, 115–121. Hoboken, NJ: John Wiley & Sons.

Belt, C. 2014. Challenges in Melt Furnace Tests. *JOM*, 66(9):1612–1615.

Belt, C. 2015. Current State of Aluminum Melting and Holding Furnaces in Industry. *JOM*, 67(11):2690–2695.

Belt, C. 2016. Case Study on Round-Top Fire Rates. In *Light Metals 2016*, ed. E. Williams, 749–751. Hoboken, NJ: John Wiley & Sons.

Belt, C., B. Golchert, P. King, R. Peterson, and J. Tessandori. 2006. Industrial Application of DOE Energy Savings Technologies to Aluminum Melting. *Light Metals 2006*, ed. T. Galloway, 881–885. Warrendale, PA: TMS.

Belt, C., and R. Peterson. 2008. Analysis of Energy Efficiency for Industrial Processes. *EPD Congress 2008*, ed. S. Howard, 53–60. Warrendale, PA: TMS.

Belt, C., R. Peterson, and D. Bequette. 2010. Five Low Cost Methods to Improve Energy Efficiency on Reverberatory Furnaces. *Energy Technology 2010: Conservation, Greenhouse Gas Reduction and Management, Alternative Energy Sources*, ed. N. Neelameggham, 71–79, Warrendale, PA: TMS.

Boeing. 2014. 2014 Environment Report. The Boeing Company. http://www.boeing.com/aboutus/environment/environment_report_14/1.2_environmental_policy_strategy.html.

Boltz, Y., and E. Boltz. 2015. Thermal Process Optimization. *Aluminum International Today*, 27(6):16–17.

Bowen, B., and M. Irwin. 2008. Coal Characteristics. Indiana Center for Coal Technology Research presentation.

Davisco. 2016. Davisco Foods International, Inc. http://www.daviscofoods.com/about/energy.html.

DOE AMO (U.S. Department of Energy Advanced Manufacturing Office). 2015. Bandwidth Study on Energy Use and Potential Energy Saving Opportunities in U.S. Iron and Steel Manufacturing.

DOE AMO (U.S. Department of Energy Advanced Manufacturing Office). 2016. Bandwidth Study on Energy Use and Potential Energy Savings Opportunities in the Manufacturing of Lightweight Materials: Aluminum.

DOE AMO and IHEA (U.S. Department of Energy Advanced Manufacturing Office and the Industrial Equipment Association). 2015. *Improving Process Heating System Performance: A Sourcebook for Industry*, 3rd ed.

EIA (U.S. Energy Information Administration). 2016. Daily Henry Hub Natural Gas Spot Price. http://tonto.eia.gov/dnav/ng/hist/mgwhhdd.htm.

ENERGY STAR. 2016a. www.energystar.gov/buildings/about-us/how-can-we-help-you/build-energy-program/guidelines.

ENERGY STAR. 2016b. Energy Efficiency and Cost Saving Opportunities for Metal Casting. Document Number 430-R-16-00.

ENERGY STAR. 2016c. Energy Star Portfolio Manager. www.energystar.gov/buildings/facility-owners-and-managers/existing-buildings/use-portfolio-manager.

EPA (U.S. Environmental Protection Agency). 2016. Overview of the Oil and National Gas Industry. https://www.epa.gov/natural-gas-star-program/overview-oil-and-natural-gas-industry.

Eppich, R., and R. Naranjo. 2007. Overview Implementation of Metal Casting Best Practices. ITP Metal Casting.

Lampe, M. 2016. How to Maximize Burner Efficiency. *Industrial Heating*, June: 37–39.

Peterson R., and C. Belt. 2009. Elements of an Energy Management Program. *JOM*, April: 19–24.

Schifo, J., and J. Radia. 2004. Theoretical/Best Practice Energy Use in Metal Casting Operations. U.S. Department of Energy Industrial Technologies Program.

Thekdi, A., and C. Belt. 2011. Waste Heat Reduction and Recovery Options for Metals Industry. *Energy Technology 2011: Carbon Dioxide and Other Greenhouse Gas Reduction Metallurgy and Waste Heat Recovery*, ed. N. Neelameggham, 17–24. Hoboken, NJ: John Wiley & Sons.

Therkelsen, P., A. McKane, R. Sabouni, T. Evans, and P. Scheihing. 2013. Assessing the Costs and Benefits of the Superior Energy Performance Program. U.S. Department of Energy Advanced Manufacturing Office.

Worrell, E. P. Blinde, M. Neelis, E. Blomen, and E. Masanet. 2010. Energy Efficiency Improvement and Cost Saving Opportunities for the U.S. Iron and Steel Industry. U.S. Environmental Protection Agency.

Index

Page numbers followed by f and t indicate figures and tables, respectively.

U

V

W

Y

www.ingramcontent.com/pod-product-compliance
Lightning Source LLC
LaVergne TN
LVHW011130300926
849160LV00008B/583
* 9 7 8 1 4 9 8 7 9 9 2 5 6 *